D. E Salmon

The Inspection of Meats for Animal Parasites

D. E Salmon

The Inspection of Meats for Animal Parasites

ISBN/EAN: 9783337229894

Printed in Europe, USA, Canada, Australia, Japan

Cover: Foto ©berggeist007 / pixelio.de

More available books at **www.hansebooks.com**

BULLETIN No. 19.

U. S. DEPARTMENT OF AGRICULTURE.

BUREAU OF ANIMAL INDUSTRY.

THE INSPECTION OF MEATS

FOR

ANIMAL PARASITES.

I. THE FLUKES AND TAPEWORMS OF CATTLE, SHEEP, AND SWINE, WITH SPECIAL REFERENCE TO THE INSPECTION OF MEATS.

By CH. WARDELL STILES.

II. COMPENDIUM OF THE PARASITES, ARRANGED ACCORDING TO THEIR HOSTS.

By ALBERT HASSALL.

III. BIBLIOGRAPHY OF THE MORE IMPORTANT WORKS CITED.

By ALBERT HASSALL.

Prepared under the direction of

Dr. D. E. SALMON,

CHIEF OF THE BUREAU OF ANIMAL INDUSTRY.

Issued February 8, 1898.

WASHINGTON:

GOVERNMENT PRINTING OFFICE.

1898.

LETTER OF TRANSMITTAL

U. S. Department of Agriculture,
Bureau of Animal Industry,
Washington, D. C., October 4, 1897.

Sir: I have the honor to transmit herewith, and to recommend for publication as Bulletin No. 19 of this Bureau, under the general title "The inspection of meats for animal parasites," a report on "The flukes and tapeworms of cattle, sheep, and swine, with special reference to the inspection of meats," prepared under my direction by Dr. Ch. Wardell Stiles, Zoologist of the Bureau. Appended to the report, and as valuable adjuncts thereto, are a "Compendium of the parasites" and a "Bibliography," prepared by Albert Hassall, M. R. C. V. S., of the Zoological Laboratory. Although the report is intended primarily for the use of the meat inspectors of this Bureau, it will be found of general interest to all sanitarians, since it treats of the communicability of certain parasites from animals to man, and suggests the necessary methods of prevention and treatment therefor. The publication and distribution of the bulletin will serve a useful purpose in disseminating knowledge of the precautions that are required to eradicate certain of the most important parasites affecting domesticated animals in this country—parasites which are a menace to the public health. Its early publication is desirable, as there is no work in the English language covering the subjects of which it treats.

Respectfully,

D. E. Salmon,
Chief of Bureau.

LETTER OF SUBMITTAL.

U. S. DEPARTMENT OF AGRICULTURE,
BUREAU OF ANIMAL INDUSTRY,
ZOOLOGICAL LABORATORY,
Washington, D. C., July 10, 1897.

SIR: I have the honor to submit herewith for publication a report covering "The flukes and tapeworms of cattle, sheep, and swine, with special reference to the inspection of meats," prepared by myself, and the corresponding "Compendium of the parasites" and "Bibliography." prepared by Dr. Albert Hassall.

This report is intended primarily for meat inspectors, and contains discussions of the various flukes and tapeworms which our Bureau inspectors are likely to meet with on the killing floors of the abattoirs. Technical zoological details have for the most part been omitted, the stress being placed upon the practical application of our zoological knowledge to questions of public hygiene.

The more important parasites for the American inspectors are: The Common Liver Fluke and the Large American Fluke, which are serious dangers to the live stock; Beef measles, Pork measles, and Hydatids, all of which bear an important relation to diseases in man.

I would direct especial attention to the Hydatids. Hydatid disease is at present comparatively rare in this country, and now is the time to attack it. By proper precautions at the abattoirs and slaughterhouses this dangerous parasite can be totally eradicated from the country. If these precautions are not carried out it will be only a question of time when this country will take its place with Germany and Australia in respect to the number of human lives sacrificed to a disease which has not yet gained much ground with us and can now be easily controlled.

The illustrations of this bulletin were prepared by Mr. Haines, artist of the Bureau.

Respectfully,

CH. WARDELL STILES.
Zoologist of Bureau of Animal Industry.

Dr. D. E. SALMON,
Chief of Bureau of Animal Industry.

4

CONTENTS.

Page.

I. The Flukes and Tapeworms of Cattle, Sheep, and Swine, with
Special Reference to the Inspection of Meats. By Ch.Wardell Stiles. 11-136
 Introduction.. 11-27
  General methods for the prevention of parasitic diseases.............. 14-15
  Treatment... 15
  The disposition of condemned meats.................................. 15-20
  Parasitic worms of cattle, sheep, and swine.......................... 20-27
  Flat worms (class *Plathelminthes*)...................................... 21-27
  Key to the flukes and tapeworms of cattle, sheep, and swine..... 21-27
 Flukes, or Trematodes (order *Trematoda*).................................. 27-67
  Distomes (flukes of the family *Fasciolidae*)............................. 28-64
  Hermaphroditic Distomes (flukes of the subfamily *Fasciolinae*).... 28-58
  Agamic, or Immature, Distomes (genus *Agamodistomum*)...... 28-29
   1. The Muscle Fluke of swine (*Agamodistomum suis*)...... 28-29
  Fascioles (Distomes of the genus *Fasciola*).................... 29-55
   2. The Common Liver Fluke (*Fasciola hepatica*) of cattle,
      sheep, swine, etc.................................... 29-48
    The effects of the Common Liver Fluke upon cattle,
      sheep, and swine............................... 34-47
    Abattoir inspection.................................. 47-48
    Jurisprudence.. 48
    The Common Liver Fluke in man.................... 48
    Varieties of the Common Liver Fluke................ 48
      (*a*) The Narrow Liver Fluke (*Fasciola hepatica
         angusta*) of Senegal cattle and man (?).... 48
      (*b*) The Egyptian Liver Fluke (*Fasciola hepatica
         aegyptiaca*) of buffalo and cattle.......... 48
      (*c*) The Common Liver Fluke (*Fasciola hepatica
         cariae*) of guinea pigs....................... 48
   3. The Giant Liver Fluke (*Fasciola gigantica*) of giraffes,
      cattle (?), and man (?)............................ 49
   4. The Large American Fluke (*Fasciola magna*) of cattle
      and deer.. 49-55
    Abattoir inspection................................. 55
  Dicrocoeles (Distomes of the genus *Dicrocoelium*)............ 55-58
   5. The Lancet Fluke (*Dicrocoelium lanceatum*) of cattle,
      sheep, and swine................................ 55-57
    Abattoir inspection............................... 57
   6. The Pancreatic Fluke (*Dicrocoelium pancreaticum*) of
      cattle and sheep.................................. 57-58
  Dioecious Distomes (flukes of the subfamily *Schistosominae*)...... 58-64
  Blood Flukes (Distomes of the genus *Schistosoma*)........... 58-64
   7. The Human Blood Fluke (*Schistosoma haematobium*) of
      man and cattle (?)................................ 58-60
   8. The Bovine Blood Fluke (*Schistosoma bovis*) of cattle
      and sheep.. 60
    The disease bilharziosis............................ 61-64
    Abattoir inspection................................. 64

5

6 CONTENTS.

THE FLUKES AND TAPEWORMS OF CATTLE, SHEEP, ETC.—Continued. Page.
Flukes, or Trematodes (order *Trematoda*)—Continued.
  Amphistomes (flukes of the family *Amphistomidae*)................... 64–67
    True Amphistomes (flukes of the genus *Amphistoma*)......... 64–67
      9. The Conical Fluke (*Amphistoma cervi*) of cattle and
        sheep 64–66
        Abattoir inspection......................... 66
      10. *Amphistoma explanatum* of zebu and cattle............ 67
      11. *Amphistoma bothriophorum* of zebu.................... 67
      12. *Amphistoma tuberculatum* of Indian oxen.............. 67
      13. *Gastrothylax crumenifer* of zebu..................... 67
      14. *Gastrothylax Cobboldii* of gayal..................... 67
      15. *Gastrothylax elongatum* of gayal and zebu............ 67
      16. *Gastrothylax gregarius* of cattle and Indian buffalo.... 67
      17. *Homalogaster paloniae* of gayal..................... 67
      18. *Homalogaster Poirieri* of zebu..................... 67
Tapeworms, or Cestodes (order *Cestoda*)........................ 68–136
  Family Taeniidae.................................... 68–136
    Hard-shell Tapeworms (Cestodes of the subfamily *Taeniinae*).... 70–125
    Hard-shell Tapeworms (genus *Taenia*)........ 70–125
      19. Beef Measles (*Cysticercus bovis*) of cattle, and its adult
        stage, the Unarmed, or Beef Measle, Tapeworm
        (*Taenia saginata*) of man.................. 71–89
        Beef measles........................ 75–77
        Abattoir inspection 77–83
        The Adult Tapeworm in man and methods of pre-
         venting the infection of cattle............... 83–89
        Key to the Adult tapeworms of man............. 84–86
      20. Pork Measles (*Cysticercus cellulosae*) of man and swine,
        and its adult stage, the Armed, or Pork Measle,
        Tapeworm (*Taenia solium*) of man 89–95
        Pork measles 92
        Abattoir inspection 92–94
        The adult and larval tapeworm in man 94–95
      21. The Thin, or Long, Necked Bladder Worm (*Cysticercus
        tenuicollis*) of cattle, sheep, and swine, and its
        adult stage, the Marginate Tapeworm (*Taenia
        marginata*) of dogs and wolves.................. 96–108
        Abattoir inspection 101
        The Adult tapeworms of dogs................... 101–108
        Key to the Adult tapeworms of dogs........... 101–102
      22. The Gid Bladder Worm (*Coenurus cerebralis*) of sheep
        and calves, and its adult stage, the Gid Tapeworm
        (*Taenia coenurus*) of dogs..................... 108–112
        Abattoir inspection 112
      23. The Echinococcus Hydatid (*Echinococcus polymorphus*)
        of man, cattle, sheep, swine, etc., and its adult
        stage, the Echinococcus Tapeworm (*Taenia echi-
        nococcus*) of dogs 113–125
        Hydatid disease in various animals............. 117–121
        Abattoir inspection 121–123
        The Adult Tapeworm in dogs................... 123–124
        Hydatid disease in man 124–125
  Adult tapeworms of cattle and sheep (subfamily *Anoplocephalinae*) 125–136
    Genus *Moniezia* 127–128
      24. The White Moniezia (*Moniezia alba*) of cattle and sheep 127
      25. Vogt's Moniezia (*Moniezia Vogti*) of sheep........... 127
      26. The Flat Moniezia (*Moniezia planissima*) of cattle and
        sheep 127–128

CONTENTS.

THE FLUKES AND TAPEWORMS OF CATTLE, SHEEP, ETC.—Continued.
Tapeworms, or Cestodes (order *Cestoda*)—Continued.
  Family Taeniidae—Continued.
    Adult tapeworms of cattle and sheep, etc.—Continued.
      Genus *Moniezia*—Continued.

Page.

27. Van Beneden's Moniezia (*Moniezia Benedeni*) of cattle and sheep.......... 128
28. Neumann's Moniezia (*Moniezia Neumanni*) of sheep .. 128
29. The Broad Moniezia (*Moniezia expansa*) of cattle, sheep, goats, etc......... 128
30. The Triangle Moniezia (*Moniezia trigonophora*) of sheep 128

Genus *Thysanosoma*.......... 128–130
  31. The Fringed Tapeworm (*Thysanosoma actinioides*) of sheep, deer, etc.......... 128–129
  32. Giard's Thysanosoma (*Thysanosoma Giardi*) of cattle (?), sheep, and swine (?)......... 129–130
Genus *Stilesia*.......... 130
  33. The Globipunctate Stilesia (*Stilesia globipunctata*) of cattle (?) and sheep.......... 130
  34. The Centripunctate Stilesia (*Stilesia centripunctata*) of cattle (?) and sheep.......... 130
Tapeworm disease of cattle and sheep.......... 131–136
Abattoir inspection.......... 136

II. COMPENDIUM OF THE PARASITES, ARRANGED ACCORDING TO THEIR HOSTS.
By Albert Hassall.......... 137–143
  Mammals (*Mammalia*).......... 137–143
    Primates.......... 137–138
    Carnivores (*Carnivora*).......... 138
    Rodents (*Rodentia*).......... 139
    Ungulates (*Ungulata*).......... 139–143
    Cetaceans (*Cetacea*).......... 143
    Marsupials (*Marsupialia*).......... 143
  Mollusks (*Mollusca*).......... 143
III. BIBLIOGRAPHY OF THE MORE IMPORTANT WORKS CITED. By A. Hassall. 145–150
INDEX TO TECHNICAL NAMES.......... 151–155
INDEX TO SUBJECTS.......... 157–161

ILLUSTRATIONS.

Page.

Fig. 1. The Muscle Fluke (*Agamodistomum suis*), occasionally found in the muscle of swine.......... 29
2. The Common Liver Fluke (*Fasciola hepatica*), natural size.......... 29
3. The Common Liver Fluke, enlarged to show the anatomical characters. 30
4. Egg of the Common Liver Fluke examined shortly after it was taken from the liver of a sheep.......... 31
5. Egg of the Common Liver Fluke containing a ciliated embryo (miracidium), ready to hatch out.......... 31
6. Embryo of the Common Liver Fluke boring into a snail.......... 32
7. Sporocyst of the Common Liver Fluke which has developed from the embryo, and contains germinal cells.......... 32
8. Sporocyst of the Common Liver Fluke, somewhat older than that of fig. 7, in which the germinal cells are giving rise to rediae.......... 32
9. Redia of the Common Liver Fluke, containing germinal cells which are developing into cercariae.......... 33
10. Redia of the Common Liver Fluke, with developed cercariae.......... 33
11. Free cercaria of the Common Liver Fluke, showing two suckers, intestine, large glands, and tail.......... 33

8 ILLUSTRATIONS.

Page.

Fig. 12. Portion of a grass stalk with three encapsuled cercariae of the Common Liver Fluke (*Fasciola hepatica*)............ 35

13. Isolated encysted cercaria of the Common Liver Fluke............ 35

14. Drawing from a microscopic preparation showing a hemorrhage in the parenchyma of the liver caused by the Common Liver Fluke.... 37

15. Drawing from a microscopic preparation showing the glandular hyperplasia of the mucosa of a gall duct caused by the Common Liver Fluke 38

16. Drawing from a microscopic preparation showing a fluke in the tissue of the liver............ 39

17. Tabular diagram of the occurrence of the Common Liver Fluke in cattle, sheep, and swine during different months of the year....... 41

18. *Limnaea truncatula*, natural size and enlarged............ 42

19. *Limnaea peregra*, natural size and enlarged 42

20. *Limnaea humilis*, natural size and enlarged 43

21. *Limnaea oahuensis*, natural size and enlarged 43

22. *Limnaea viator*, natural size and enlarged............ 43

23. The Narrow Liver Fluke (*Fasciola hepatica angusta*), natural size..... 48

24. The Narrow Liver Fluke, enlarged to show the anatomical characters. 49

25. The Egyptian Liver Fluke (*Fasciola hepatica aegyptiaca*), natural size. 49

26. The Egyptian Liver Fluke, enlarged to show the anatomical characters 50

27. The Giant Liver Fluke (*Fasciola gigantica*), enlarged to show the anatomy 50

28. The Large American Fluke (*Fasciola magna*), natural size............ 51

29. Macerated specimen of Large American Fluke, showing the digestive system and acetabulum............ 51

30. Macerated specimen of Large American Fluke, showing the anatomical characters............ 52

31. A section of the cuticle of Large American Fluke, showing the spines. 53

32. Egg of Large American Fluke, showing the germ cell, surrounded by a large number of vitelline cells, and an eggshell provided with a cap 53

33. Ciliated embryo (miracidium) of Large American Fluke within the eggshell............ 53

34. Free embryo (miracidium) of Large American Fluke, showing ciliated epithelium, boring papilla, rudimentary oesophagus, and intestine; eye-spots situated above the ganglionic mass, and germ cells....... 54

35. Cyst in the liver, caused by Large American Fluke............ 54

36. Lancet Fluke (*Dicrocoelium lanceatum*), natural size............ 55

37. Lancet Fluke, enlarged to show the anatomical characters 55

38. Egg of Lancet Fluke with contained embryo............ 56

39. Free embryo (miracidium) of the Lancet Fluke 56

40. The Pancreatic Fluke (*Dicrocoelium pancreaticum*), enlarged to show the anatomical characters............ 56

41. Male and female specimens of the Human Blood Fluke (*Schistosoma haematobium*), enlarged............ 57

42. Anterior portion of male Human Blood Fluke, showing the anatomical characters............ 58

43. Anterior portion of female Human Blood Fluke, showing the anatomical characters............ 59

44. Egg of Human Blood Fluke with contained embryo, passed in the urine............ 60

45. The Bovine Blood Fluke (*Schistosoma bovis*), male and female........ 60

46. Cross section of Bovine Blood Fluke, showing the position of the female in the gynaecophoric canal............ 61

47. Eggs of Bovine Blood Fluke, showing the peculiar process on the end 62

Page.

Fig. 48. Ureter of an Egyptian, with numerous uric-acid concretions, as a result of blood-fluke infection.. 63

49. Conical amphistomes (*Amphistoma cerri*) in the rumen; tubercles from which the parasites have loosened..................................... 64

50. Dorsal view of a Conical Amphistome, showing the anatomical characters 64

51. Dorsal view of the free embryo (miracidium) of the Conical Amphistome about to enter the intermediate host............................. 65

52. Sporocyst of the Conical Amphistome resulting from the transformation and development of the embryo................................... 65

53. Adult redia of the Conical Amphistome of the first generation, thirty-nine days after the infection of the intermediate host with embryos. 66

54. Young redia of the Conical Amphistome of the second generation in which the cercariae develop... 66

55. Mature cercaria of the Conical Amphistome, the stage which gains access to cattle and sheep... 67

56. *Amphistoma bothriophorum*.. 67

57. Enlarged dorsal view of *Gastrothylax crumenifer*........................ 68

58. Enlarged ventral view of *Gastrothylax crumenifer*....................... 68

59. Enlarged view of anterior extremity of *Gastrothylax crumenifer*...... 68

60. Enlarged view of posterior extremity of *Gastrothylax crumenifer*. 68

61. Enlarged view of *Gastrothylar crumenifer*, with ventral pouch open... 69

62. Dorsal view of *Gastrothylax crumenifer*, magnified to show the anatomical characters... 69

63. *Gastrothylax Cobboldii*, lateral view.. 69

64. *Gastrothylax elongatum*.. 70

65. Dorsal view of *Gastrothylax gregarius*.. 71

66. Lateral view of *Gastrothylax gregarius*....................................... 71

67. *Homalogaster paloniae*, ventral view.. 72

68. Section of a beef tongue heavily infested with beef measles.......... 72

69. Several portions of an adult Beef-measle Tapeworm (*Taenia saginata*) from man, showing the head on the anterior end and the gradual increase in size of the segments... 73

70. Dorsal, apex, and lateral views of the head of Beef-measle Tapeworm, showing a depression in the center of the apex 74

71. Segments from various strobilae of Beef-measle Tapeworm, showing forms of proglottids which are occasionally found.................. 75

72. Sexually mature segment of Beef-measle Tapeworm...................... 76

73. Gravid segment of Beef-measle Tapeworm, showing lateral branches of the uterus... 77

74. Egg of Beef-measle Tapeworm, with thick eggshell (embryophore), containing the six-hooked embryo (oncosphere) 81

75. A piece of pork heavily infested with pork measles (*Cysticercus cellulosae*).. 90

76. An isolated Pork-measle Bladder Worm with extended head.......... 90

77. Several portions of an adult Pork-measle Tapeworm (*Taenia solium*). 91

78. Large and small hooks of Pork-measle Tapeworm 92

79. Mature sexual segments of Pork-measle Tapeworm, showing the divided ovary on the pore side... 92

80. Segment of Pork-measle Tapeworm in which the uterus is about half developed 92

81. Gravid segment of Pork-measle Tapeworm, showing the lateral branches of the uterus.. 94

82. Eggs of Pork-measle Tapeworm.. 94

83. Half of hog, showing the portions most likely to become infested with measles.. 96

Page.

FIG. 84. The Thin-necked Bladder Worm (*Cysticercus tenuicollis*) with head extruded from body, from cavity of a steer 97

85. The Marginate Tapeworm (*Taenia marginata*) 97

86. Head of the Marginate Tapeworm 98

87. Small and large hooks of *Taenia marginata*, *T. serrata*, and *T. coenurus* 98

88. Sexually mature segment of the Marginate Tapeworm 99

89. Gravid segments, showing the lateral branches of the uteri of *Taenia serrata*, *T. marginata*, and *T. coenurus* 99

90. Egg of the Marginate Tapeworm, with six-hooked embryo 100

91. Portion of the liver of a lamb which died nine days after feeding with eggs of the Marginate Tapeworm, with numerous "scars," due to young parasites 100

92. Cross section of the liver of a lamb which died nine days after feeding with eggs of the Marginate Tapeworm 101

93. Young cysticerci of the Marginate Tapeworm 103

94. Skull of a sheep showing the brain infested with a Gid Bladder Worm (*Coenurus cerebralis*) 106

95. An adult Gid Tapeworm (*Taenia coenurus*) 107

96. Sexually mature segment of the Gid Tapeworm 108

97. Brain of a lamb infested with young Gid Bladder worms 108

98. Sheep's skull, the hind portion thin and perforated, due to the presence of Gid Bladder worms 109

99. An isolated Gid Bladder Worm, showing the heads 110

100. Diagrammatic section of a Gid Bladder Worm 111

101. Portion of hog's liver infested with Echinococcus hydatid 112

102. Portion of the intestine of a dog infested with the adult Hydatid Tapeworm (*Taenia echinococcus*) 114

103. Adult Hydatid Tapeworm, enlarged 115

104. Hooks of Adult Hydatid Tapeworm 115

105. Diagram of an Echinococcus hydatid 116

106. A racemose Echinococcus 117

107. Section through a multilocular Echinococcus 117

108. A multilocular Echinococcus from the liver of a steer 118

109. A multilocular Echinococcus from the pleura of a hog 118

110. Lymphatics of a steer infested with the so-called "Tongue worm" (*Linguatula rhinaria*) 119

111. Portions of an adult Flat Moniezia (*Moniezia planissima*) 120

112. Three views of heads of the Flat Moniezia 121

113. Dorsal view of sexually mature segment of the Flat Moniezia 122

114. Dorsal view of gravid segments of the Flat Moniezia, showing the uterus 126

115. Egg of the Flat Moniezia 126

116. Portions of an adult specimen of the Broad Moniezia (*Moniezia expansa*) 127

117. Three views of the head of the Broad Moniezia 129

118. Sexually mature segments of the Broad Moniezia 129

119. Gravid segment of the Broad Moniezia 130

120. Portions of an adult specimen of the Triangle Moniezia (*Moniezia trigonophora*) 131

121. Sexually mature segments of the Triangle Moniezia 132

122. Adult specimen of the Fringed Tapeworm (*Thysanosoma actinioides*) ... 133

123. Ventral and apex views of the head of the Fringed Tapeworm 134

124. Segments of the Fringed Tapeworm, showing canals and nerves, fringed border, testicles, and uterus 136

THE INSPECTION OF MEATS FOR ANIMAL PARASITES.

I. THE FLUKES AND TAPEWORMS OF CATTLE, SHEEP, AND SWINE, WITH SPECIAL REFERENCE TO THE INSPECTION OF MEATS.

By CH. WARDELL STILES, Ph. D.,

Zoologist of the Bureau of Animal Industry.

INTRODUCTION.

The object of the report.—The present report on "The flukes and tapeworms of cattle, sheep, and swine, with special reference to the inspection of meats," is intended primarily for the use of meat inspectors, and an effort has been made to bring together in systematic order the more important facts relating to the flukes and tapeworms which inspectors are likely to find in the abattoirs and slaughterhouses. For several reasons it is important that meat inspectors should be well informed upon both the practical and the theoretical considerations of this subject:

First. Since certain parasites (*Cysticercus cellulosae* and *C. bovis*) are directly transmissible to man through the use of meat, a knowledge of these worms will enable inspectors to prevent the spread of their tapeworm stage among human beings by condemning the infested meat or subjecting it to processes which will render it harmless. The rigid system of meat inspection in Germany has resulted in an actual decrease in tapeworm disease (by *Taenia solium* and probably also by *T. saginata*) in man and in the frequency of *C. cellulosae* in the human eye.

Second. Condemnation and destruction of organs infested with certain other parasites (*Echinococcus, Coenurus, Cysticercus tenuicollis*) will prevent the spread of these parasites in their tapeworm stage to dogs, and by that means prevent the reinfection of man (by *Echinococcus*) and of domesticated animals (by *Echinococcus, Coenurus, Cysticercus tenuicollis*); in this case prevention of tapeworm disease in dogs, though of comparatively little importance so far as the dogs are concerned, becomes very important not only in public hygiene (in the prevention

11

of disease in man and animals), but also from an economic standpoint, preventing financial loss to stock raisers from disease and death in their herds and flocks caused by these worms. The destruction of livers heavily infested with flukes will also result indirectly in decreasing fluke disease in man and live stock.

Third. Certain animal parasites (*Cysticercus cellulosae*, *C. bovis*, *Echinococcus*, etc.) may under certain conditions bring about pathological appearances in the meat which may at first sight be mistaken for tuberculosis. It is hardly necessary to insist upon the importance of a differential diagnosis between tuberculosis and diseases caused by animal parasites.

It is thus seen that the meat inspector is destined to render an important public service in the prevention of parasitic diseases, not only in man, but among domesticated animals.

Secondarily, this report is intended for the stock raiser, and an attempt has been made in the text to give him such information regarding the various parasites discussed as will be useful in preventing the spread of parasitic diseases among his animals. The stock raiser, whether the owner of a large herd or of but one or two animals, should never lose sight of the fact that his stock is raised not only as a money investment for himself, but as food for his fellow-men. To allow the introduction of certain diseases among his animals means not only a financial loss to himself, but a loss of health or life to those who may use these animals for food. To prevent these diseases is to increase the value of his investment and to aid the health authorities in preventing disease among his neighbors and his neighbors' stock. The stock raiser's position is therefore based not only upon dollars and cents, but also upon the broader plane of ethics, and he who intentionally or unintentionally and persistently loses sight of the ethical side of his occupation must necessarily suffer from the financial standpoint. The unprincipled action of placing diseased live stock on the market, instances of which can be cited from all civilized countries, is indeed a very short-sighted policy, which will sooner or later tell upon the purse of him who descends to such action, as well as upon the health of the community. A person who aids in concealing a smallpox or diphtheria patient from the health authorities and thus jeopardizes the health of his friends and neighbors justly earns the contempt of his fellow-beings as well as the punishment provided by law in some places; and a person who knowingly places diseased live stock or diseased meat on the market and thus endangers the health of those who consume the meat is none the less worthy of contempt and punishment.

In the third instance, this report is intended for butchers who handle meat which has not been inspected, as is unfortunately the case in many places, particularly in smaller towns. The writer has personally seen many towns where the meat supply was drawn almost entirely from local slaughterhouses, in which there was no inspection. Frequently the butchers raised their own stock, or a portion of it, on the grounds of the slaughterhouses and under unhygienic conditions, which were not only most favorable to the spread of disease, but which must necessarily have resulted in the spread of infection among the animals raised upon the premises and in the neighborhood; especially in cases where the slaughterhouse was located on the banks of a creek or river.

In writing a report for the information of these three classes of persons, the author is well aware that technical language is not desired by the stock raiser and the butcher; a considerable amount of technical detail is, however, necessary in treating this subject in a manner which will be exact and complete enough for the expert inspector, who must view the questions from different standpoints. These technical details consist chiefly (1) of the classification and analytical keys to the various worms, necessary in order to properly determine the parasites found; (2) of detailed synonymy of each form, necessary because so many of the parasites are described in various works under different names; (3) details in regard to the life history of the worms, necessary in order to establish the proper methods of prevention; and, (4) details in the pathological appearances of diseased organs, necessary in order to differentiate between diseases which may bear a close resemblance to one another.

This technical discussion, necessary as it is to the expert inspector, has been forced to the background as much as possible by placing it in small type or in footnotes, and any, except sanitary officers and zoologists, who read this bulletin will do well to rely chiefly upon the discussion in large type.

Scientific nomenclature and synonymy.—One of the greatest aids in scientific work, giving exactness to statements and rendering the names of animals and plants international, is the use of Latin names for all 'plants and animals. These names should be given according to certain regulations agreed upon by workers in science, but owing to the disregard of these rules by some authors, many of the parasites discussed have received numerous technical names. In this paper the writer has endeavored to follow the international rules in selecting the technical name used for each parasite, and this name alone should be quoted in referring to the worms. The lists of synonyms are intended only as tables of reference, in order to trace the parasites as described by different authors.

Authorities consulted.—The majority of the parasites mentioned in the report have been known for many years and much has already been published upon them. In writing the report, therefore, I have not only drawn from my own personal studies, but have not hesitated to use the entire literature at my disposal. A list of the chief works consulted is given on pages 145–150, and of these I have used with special freedom Zürn (1882), Blanchard (1885–95), Neumann (1892), Railliet (1893), Ostertag (1895), and my own papers.

GENERAL METHODS FOR THE PREVENTION OF PARASITIC DISEASES.

"A well-regulated system of slaughterhouses is as necessary to public health as is a well-regulated system of schools to public education."

Under the subject of prevention there will here be considered chiefly those rules which apply to the parasites discussed in this paper. The methods of prevention naturally fall under several heads:

(1) *Segregation of slaughterhouses.*[1]—The first and most important step to be taken in order to prevent the spread of parasitic diseases is to segregate the slaughterhouses. In many places, especially in the West, we find two, three, four, or even five small slaughterhouses on the outskirts of a town of 300 to 2,500 inhabitants. These slaughtering places are scattered north, east, south, and west of the town; as they are often outside of the corporation limits, they do not come under the direct control of the local board of health; few, if any, of the State boards pay any attention to them, and as a result the meat supply is often without sanitary supervision.

The general rule may be laid down that every slaughterhouse is a center of infection for the surrounding neighborhood, not only of diseases caused by animal parasites, but also of other diseases, such as hog cholera, swine plague, tuberculosis, etc. The first step to be taken, therefore, is to reduce the number of localities from which infection may spread, and there is evidently only one way to do this, namely, to compel all the butchers of a town to do all of their killing at the same slaughterhouse. If the slaughtering is all done at one place, it is comparatively easy to control the class of animals used; but when numerous slaughterhouses exist, it is practically impossible to supervise the premises.

In many European cities and towns the slaughterhouse is built either at municipal expense or by a stock company, and stalls are let to the butchers for killing purposes. This plan has been found very satisfactory.

The places of slaughtering should be built of some more durable material, as brick, rather than wood; the less wood used the easier it is to keep the place clean. Even the floors should, if possible, be of brick, stone, or asphalt.

(2) *Sanitary supervision of slaughterhouses.*—There should be a competent veterinary inspector appointed as director of every slaughterhouse, with assistants if necessary. It should be the duty of the director and his assistants to see that the stalls and grounds are kept in proper sanitary condition, and that the offal is properly disposed of. In small towns, where there is not enough offal to pay for preparing it as fertilizer, there seems to be no valid sanitary objection to feeding the offal of healthy cattle and sheep to hogs; but offal of hogs should under no circumstances be fed to other hogs, unless it is first thoroughly cooked.

---

[1] Cf. Stiles, 1897. The Country Slaughterhouse as a Factor in the Spread of Disease. (Yearbook of the U. S. Department of Agriculture for 1896, pp. 155-166.)

(3) *Meat inspection.*—There should be a regular inspection, by a competent veterinarian, of all meats before they are allowed to leave the slaughterhouse.

(4) *Dogs and rats.*—Dogs should be excluded from slaughterhouses and meat shops, and all stray and ownerless dogs should be killed. This will prevent the spread of a number of dangerous parasites. Rats are common factors in spreading diseases from slaughterhouses, although they do not come into consideration in connection with any of the parasites discussed in this report.

(5) *The raising of hogs and other animals at slaughterhouses* is a custom which can not be too severely condemned, and the farmer who grants to a butcher the privilege of slaughtering on his farm in exchange for the use of the offal as feed simply bids for disease.

(6) *Deserted premises.*—In the segregation of slaughterhouses, which must come sooner or later, care should be taken to properly dispose of the houses which are deserted; an attempt should be made to kill the rats on the deserted premises, in order to prevent their spreading disease by wandering to neighboring farms, etc.

(7) *Domesticated animals* must not be allowed access to human excreta or to water supply contaminated by drainage from privies, vaults. etc.

TREATMENT.

The treatment of the verminous diseases of cattle, sheep, and swine, discussed in this report, may be summed up in two rules:

(1) The treatment for the larval tapeworms and the liver flukes must be preventive, as no medicinal treatment known is satisfactory.

(2) The treatment for the adult tapeworms and the intestinal flukes should be medicinal, as this is effective, and the life history of most of these worms still being problematical, we have no satisfactory data upon which to base preventive measures.

(For details regarding prevention and treatment,[1] see these captions under each parasite.)

THE DISPOSITION OF CONDEMNED MEATS.

The proposition that diseased meats which are dangerous as articles of food should not be allowed on the market is one which will receive universal support from all sanitarians and also from the thinking pub-

---

[1] In connection with the subject of treatment, I would call the attention of veterinarians to the necessity of not forgetting that a prescription written in one country does not mean the same in all countries. In dealing with pounds, ounces, and grains the apothecaries' weight, United States, agrees with the imperial standard troy, but many of the articles used in dosing large herds are purchased at avoirdupois weight. The apothecaries', United States, and the imperial liquid measures do not agree, a point which should be borne in mind in utilizing English formulae in this country. Have not many accidents occurred because English formulae were taken, and the fact overlooked that the English gallon is one-fifth larger than the United States gallon? Hutcheon's wireworm treatment (pp. 133–135), if adopted in this country without making due allowance for the difference in the size of the gallon, would probably result in heavy losses to the sheep owner.

lic. The question, however, arises as to the classes of diseased meats and the stages in these diseases that justify their condemnation or that justify their sale and the method of their disposal if condemned. It is not the purpose of this report to discuss the general aspect of these questions, but only to discuss them so far as the diseases caused by animal parasites are concerned.

In some foreign cities regulations exist, or have existed, compelling the burial or burning of meats affected with certain parasitic diseases. To such extreme regulations we are opposed for several reasons. In the first place, such destruction by burial or burning is in itself an expense. It also results in a total and unnecessary loss of the carcass. Again, the burial of a diseased carcass, unless buried in quicklime or other destructive material, does not meet either the practical or the theoretical requirements of destruction of diseased material. Take the disease trichinosis, for instance. In some places the carcasses of trichinous hogs have been buried by order of the sanitary officials. After this has been done, the owners of the carcass have disinterred the hog and it has been used for food![1] Even had these men not disinterred the body and fed it to their friends and customers, the grave would have been accessible to rodents, such as rats, field mice, etc., which would be likely to feed upon the carcass, and thus become infected with the disease, resulting in a possible (theoretical!) ultimate transmission of the disease to other hogs. Finally, the writer is opposed to this method of destruction (?) on the ground that diseased or partially diseased carcasses can be utilized under certain conditions and restrictions, so that the owner will not lose the entire amount of his investment.

Three methods in particular are open, the method selected being dependent (1) upon the nature, extent, or stage of the disease, and (2) the facilities at hand. These methods are: (1) Utilization as fertilizer; (2) rendering the meats harmless by cold storage, cooking, or preserving, and then placing them upon the market; (3) selling the meats under a declaration of their character.

In determining the extent or stage of the disease and its relation to the method of disposition of the carcass, the opinion of the meat inspector must, of course, be based upon certain general principles and must naturally be final.

Utilization as fertilizer.—There is no parasitic disease known which will withstand the degree of heat used at the large abattoirs in the preparation of fertilizers. "Tanking for fertilizers" is therefore an absolutely safe method for the disposition of condemned meats, no matter how serious the infection is or to what extent the disease has progressed.

In connection with some parasitic diseases, however, a question arises as to the necessity of condemning to the tank certain diseased con-

---

[1]This has happened a number of times in Germany, one case being reported within less than a year! (See Zeitschr. f. Fleisch- und Milchhygiene, 1897, VII, (5), p. 104.)

ditions. A case of generalized cestode-tuberculosis (*Cysticercus boris*) should undoubtedly be "tanked," but in a very light infection the question takes a different aspect, namely: Can not the diseased portion be cut out and the rest of the carcass be placed on the block? To allow such meat on the market, leaving the consumer to suppose that he is purchasing a first-class article, is evidently an injustice to the buyer, for it is by no means certain that all of the parasites have been detected and removed. To condemn a light infection of this disease is, on the contrary, an injustice to the dealer, for there are methods by which the remaining parasites, if any, may be rendered harmless, and in this case the dealer could be saved a part of his loss. To judge between those cases in which the carcass is absolutely unfit for food, and therefore to be condemned, and those cases in which the carcass may be treated according to methods which will destroy the remaining but undiscovered parasites, thus rendering the meat fit for food, is a point upon which the expert meat inspector must decide.

To follow up the example cited, let us examine the effects of cold storage, cooking, and salting. It is evident that the method chosen must depend upon the facilities at hand. At a large abattoir any of these methods might be followed, but at a small country slaughter-house the choice would be restricted.

Cold storage.—Experiment shows that the parasite under discussion (*Cysticercus boris*) dies about two to three weeks after the death of its host. Three weeks of cold storage would therefore render a light infection of this kind absolutely harmless, and the meat could safely be placed on the block. With the disease known as pork measles the parasites live for a month or more, so that more care would be necessary in dealing with it.

Cooking.—Many of the abattoirs voluntarily tank for canning certain meats of inferior quality. The heat to which these meats are subjected is not so great as that used in tanking for fertilizers, but as *Cysticercus bovis* can not survive a temperature of 140° F. (see p. 81) for five minutes, and as the meats tanked for canning are thoroughly cooked, it may safely be asserted that a light case of "beef measles" would be rendered perfectly harmless by the cooking preparatory to canning.

The same applies to cases of trichinosis. The parasite of this disease can not withstand a heat of 70° C. (=158° F.), so that if trichinous pork is cooked until the entire piece has reached this temperature and assumed a light-gray color, the disease is rendered nontransmissible to man.

Salting.—The parasite of "beef measles" is killed in twenty-four hours by the action of salt solution, and we have found no case where the parasite of trichinosis has been able to withstand four months in the "pickling vats." In both of these cases it must be remembered that it takes some time for the salt to thoroughly permeate the tissue. It

would accordingly not be safe to assume that in a piece of measly beef
which had been placed in brine for twenty-four hours the parasites had
been killed. The length of time necessary to guaranty the result is, of
course, dependent upon the size of the piece of meat.

Selling infected meats under declaration.—While the large abattoirs
have means at their command by which cases of light infection may be
rendered noninfectious, the smaller slaughterhouses are at more of a
disadvantage in this respect. Cooking and salting would be possible
for some—perhaps all of them—while cold storage would often be out
of the question.

In this connection, it will be interesting to study for a moment a system which is quite extended in certain parts of Europe. Reference is
made to the German "*Freibank*" or "*Finnenbank*." Under this system
certain meats of inferior quality are allowed to be placed on the market
under given conditions. One of these conditions is that they must be
sold in a specified meat stall or counter, known as the "*Freibank*" or
"*Finnenbank*," where the true nature of the meat must be made known
to the purchaser. Naturally, such meats are sold at a lower price than
the meats offered in open market, thus enabling many of the poorer
classes to purchase meat who can not afford to pay the regular prices.
Meats which are absolutely dangerous from a sanitary standpoint are,
of course, excluded from these special meat counters, and in some
instances the law requires that even these meats of inferior quality,
which are harmful in some cases, though not dangerous, must be rendered harmless before being sold.

In the United States inspected meats are, generally speaking, either
passed and allowed to go upon the open market or condemned and thus
excluded from the market. The German system of the "*Freibank*" practically results in dividing the meats into three[1] classes, namely, first,
meats which may be sold in open market—good or first-class meats

---

[1] Strictly speaking, the Imperial German law of May 14, 1879, divides meats into
five classes, as follows:

"1. Good or first-class wares which may be placed upon the open market without
restrictions. This corresponds to the '*bankwürdiges Fleisch*' of the South German
meat inspection regulations.

"2. Meat which may be placed upon the market under declaration and sold
as 'spoiled (or waste) goods in the sense of the food laws.' Other disposition
of this meat (as use in one's own family or presentation to other persons), is not
prevented by law. This meat is called '*nichtbankwürdiges Fleisch*' in the older
regulations.

"3. Meat which is unconditionally dangerous or injurious to health, the use of
which, under any condition, as food for man, even use in one's own family, presentation to other people, or permitting it to be taken away, etc., is forbidden by law.
This meat must be disposed of in such a way as to render it harmless.

"4. Meat which is injurious to health under certain conditions, but which can be
rendered harmless by proper manipulation, such as cooking, sterilizing, pickling,
etc. After the meat has been rendered harmless it may be placed on the market as

("*gute oder tadellose Ware*," of North Germany, "*bankwürdiges Fleisch*," of South Germany, also called "*bankmässig*" or "*ladenrein*"); a second class of meats which may be sold only under declaration of their true character, in many cases only after having been cooked or salted under official supervision ("*nichtbankwürdig*," "*nichtbankmässig*," "*nichtladen-rein*"); a third class of meats which are unconditionally condemned, and therefore excluded from the market.

History of the "Freibank."—The system of the German "*Freibank*" and compulsory declaration of the condition of inferior meats is very old. The municipal laws of Augsburg in 1276 prescribed that inferior meat should not be sold without giving notice as to its quality. In 1404 the municipal laws of Wimpfen provided that the "*Freibank*" (from the German "*frei*," free, here in the sense of unconnected or separated, and "*Bank*," a counter or stall) should be situated three paces away from the regular counters. The "*Freibank*" (free stall) was, therefore, one which was free or separate from the regular counters. The term "*Finnenbank*" is sometimes used for these special meat stalls because the measly meat ("*finniges Fleisch*") especially is sold at these places. This system of "*Freibank*" has been extended to most of the slaughterhouses of Germany, and is rapidly extending in France, Belgium, and Italy.

The economic importance of the system is seen from the following statistics taken from Ostertag:

In the Kingdom of Saxony in 1892, 0.25 per cent of the animals

---

spoiled (or waste) meat, in the sense of the food law. In regard to selling this kind of meat raw compare the legal decisions:

" An explicit statement by the seller that the meat, which is rendered harmless by cooking, is to be eaten only when cooked protects the merchant from penalty." (Urt. IV, v. II, 7, 1884.)

"A simple statement regarding the unwholesome condition of the meat on the part of the merchant to the purchaser does not, however, render the former free from penalty, for the danger to the communal interests of the act is not thereby obviated." (Urt., v. 15, 1 and 29, 9, 1885.)

"5. Finally, there should be recognized meat which is spoiled beyond use [literally, spoiled in high degree], i. e., meats which, though not unwholesome, have lost their value as food for man because of extensive changes in the tissue (for example, watery meat, meat and organs which are heavily infested with parasites, etc.). Such meats are to be judged as 'unfit for food,' and can be looked upon as 'spoiled' in the sense of sec. 367 of the penal code, and offering for sale and selling such meats are plainly forbidden by this paragraph. Their use in the household of the owner can not, however, be forbidden on grounds of the Imperial regulations. In order to prevent underhand traffic with such meat, it is provided that meat which is spoiled beyond use is to be entirely excluded from the market, except in such cases as portions of the same, such as the fat in heavily infected cases of pork measles, can be used for food.

" In meat wares we further distinguish imitations [*nachgemachte*] (meats which are treated in such a way as to appear different from what they really are, Urt. I, v. 15, 5, 1882), and adulterations [*verfälschte*] (meats which do not possess those qual_ities which they are supposed to possess in reliable traffic).

"I will call attention to the fact that the expert must use the word 'spoiled' [*verdorben*] only in the legal sense and not in the sense of decomposed meat, for decomposing [*faulende*] meat is injurious to health."—OSTERTAG, 1895, pp. 100, 101, et al..

slaughtered for food were unconditionally condemned, while 0.42 per cent of the animals slaughtered were sold at the *"Freibank."*

In Leipzig during 1891 the meat of 604 cattle, 89 calves, 28 sheep, 983 hogs, and 104 pieces, representing a total weight of 271,608 kilograms (about 543,216 pounds), was used by the *"Freibank."*

The average receipts per pound for the first quality meats and for the meat sold at the *"Freibank"* after deducting fees were as follows:

First quality.	Price.
Beef...........................	57.6 pfennige, about $0.144
Veal...........................	55.5 pfennige, about .138½
Mutton.......................	58.8 pfennige, about .147
Pork...........................	61.0 pfennige, about .152½

Freibank.	Price.
Beef...........................	53.8 pfennige, about $0.134½
Veal...........................	44.2 pfennige, about .110½
Mutton.......................	54.5 pfennige, about .136½
Pork...........................	57.4 pfennige, about .143½

Ostertag (1896) has recently published a detailed compilation giving the data concerning the sale of measly beef in 38 cities in Germany. At first there was great prejudice against the meat, so that in some cases the price fell to 2½ cents per pound; but as this prejudice wore off the price went up 6, 8, and 10 cents per pound. In some places the demand for this cheaper meat is greater than the supply.

Objections to the *"Freibank"* have been raised by some parties, but we are unable to see wherein this system is unfair either to the dealer or to the purchaser, for no one is obliged to buy this meat who does not wish to do so, while anyone who wishes a cheaper class of meat can purchase it at the *"Freibank"* with the full knowledge of the condition of the meat he is buying. It is perfectly safe to use the meat when thoroughly cooked, and the dealer is able to economize in his business. We take the decided stand, however, that it is far better to subject all of these meats to thorough cooking or other methods of safeguarding before they are placed upon the market.

PARASITIC WORMS OF CATTLE, SHEEP, AND SWINE.

The term "cattle" in this report is used in the American sense of the word, i. e., for the species known zoologically as *Bos taurus*, the only bovine animal at present slaughtered in this country. Other animals also are known under the terms cattle, bulls, etc., in some countries, and a few parasites found in these animals are mentioned briefly in this report. These parasites are cited because the same species, or at least the same genera, are likely to infest *Bos taurus* sooner or later.

The term "sheep," as used here, refers to the only species of sheep slaughtered in the United States, i. e., *Ovis aries.*

The terms "hog," "pig," and "swine" refer to the only species of domesticated swine found in this country, i. e., *Sus scrofa domestica.*

The parasitic worms found in cattle, sheep, and hogs belong to two different zoological groups, known as Flat worms (*Plathelminthes*) and

Round worms (*Nemathelminthes*). The Flat worms alone are discussed in this report.

FLAT WORMS (Class *Plathelminthes*).

The Flat worms include at present five orders, only two of which, namely, the flukes (*Trematoda*) and the tapeworms (*Cestoda*), are discussed in this report.

FLUKES, OR TREMATODES.—The flukes found in cattle, sheep, and swine vary in size from a few lines to 4 inches in length and from one or more lines to an inch or more in breadth. They are found in the liver, lungs, intestine, and body cavity, and occasionally in other parts of the body. None of the species found in cattle, sheep, or swine are directly transmissible from these animals to man, although three of the species occasionally infest man. At least two of the species render the organs in which they occur unfit for food when present in numbers; they also injure the animals to a greater or less degree, although the extent of injury in cattle has possibly been overestimated; one form is particularly injurious to sheep.

TAPEWORMS, OR CESTODES.—Cestodes occur as larval forms (*bladder worms*) or as adult forms (*tapeworms, strobilae*).

Larval tapeworms.—The larvæ, or bladder worms (*Cysticercus, Coenurus, Echinococcus*), are found in the liver, lungs, brain, muscles, or other organs except the intestinal tract, and do not reach maturity until they are transmitted to meat-eating animals. The most important bladder worms considered in this report are: (1) The *Beef-measle Bladder Worm*, and (2) the *Pork-measle Bladder Worm*, both of which develop into tapeworms in man; (3) the *Gid Bladder Worm*, which causes gid, or turnsick, in sheep; and, (4) the *Hydatid*, which causes hydatid disease in man and various domesticated and wild animals. When eaten by dogs the two latter bladder worms develop into Adult tapeworms.

Adult tapeworms.—Several different species are found in the intestine of cattle and sheep. They injure their hosts, but are not transmissible to man in any stage of their development.

The following key will aid the reader in determining the various flukes and tapeworms discussed in this report. A certain amount of technical knowledge is valuable in the use of this key, which is based upon zoological characters. Some liberty has, however, been taken with the anatomical characters in order to make the key as simple as possible; and it is believed that most, if not all, of the forms mentioned can be more or less definitely determined by comparing the key, especially the *habitat* given for each form, with the figures of the parasites, even if one is unable to follow the more technical statements.

KEY TO THE FLUKES AND TAPEWORMS OF CATTLE, SHEEP, AND SWINE.

[For the species thus far positively known to have been found in North America, follow Roman type. As the characters given are confined to the forms discussed in this report, this key should not be relied upon to classify the parasites of other animals.]

(1) Parasitic in the liver, lungs, pancreas, veins, abdominal cavity, or intestine of cattle, sheep, and swine, more rarely encysted in muscles of swine. Unsegmented Flat worms; intestinal tube present; anus absent; mouth with one sucker .... *Flukes,* 2.

Parasitic in the intestine, more rarely in bile ducts, as adult segmented worms; or in the liver, lungs, muscles, etc., as unsegmented bladder worms. Intestinal tube absent; head with four suckers........................... *Tapeworms*, 14.

FLUKES.

(Trematoda.)

(2) Parasitic in liver, lungs, pancreas, veins, or abdominal cavity of cattle, sheep, or swine, rarely encysted in muscles of swine. Ventral sucker (acetabulum) on anterior half of body .... *Fasciolidae*, 3.
Parasitic in intestinal tract or gall ducts of sheep, cattle, zebu, gayal, or buffalo. Ventral sucker (acetabulum) at posterior extremity .... *Amphistomidae*, 9.

Fasciolidae.

(3) Parasitic in liver, lungs, pancreas, or abdominal cavity of cattle, sheep, or swine, more rarely encysted in muscles of swine. Hermaphrodites.... *Fasciolinae*, 4.
Parasitic in the blood of cattle and sheep; eggs found forming egg tumors in genito-urinary tract or colon. Sexes separate .... *Schistosominae*, 8.

Fasciolinae.

(4) Parasitic in liver, lungs, pancreas, or abdominal cavity. Mature flukes, or forms in which the genital organs are developed to an extent which permits of a determination of the genus. Acetabulum sessile; genital pore between oral sucker and acetabulum; oral sucker unarmed.............................. 5.
Encysted in muscles of swine, very rare. Must not be mistaken for trichinae. Immature fluke, in which the organs do not permit of a determination of the genus. Body (fig. 1) 0.5 mm. long, elliptical, grayish, transparent; oral sucker terminal; ventral sucker near the middle of the body; pharynx followed by a short oesophagus and two simple intestinal caeca, which extend slightly beyond the middle of the body; in front of acetabulum are four large unicellular glands with rather long ducts, extending to oral sucker; three primordial genital glands in distal half of body; terminal excretory canal median, branching immediately distal of testicles.
The Muscle Fluke of Swine (*Agamodistomum suis*), p. 28.
(5) Parasitic in liver, lungs, rarely abdominal cavity of cattle, sheep, or swine. Body large, shaped like a flat fish, dark colored; intestinal caeca, testicles and ovary profusely branched; freshly laid egg does not contain embryo.
Fascioles (*Fasciola*), 6.
Parasitic in liver or pancreas of cattle, sheep, or swine. Body smaller; intestinal caeca very simple, long, tubular, extending beyond acetabulum to posterior portion of body; oesophagus comparatively short; genital pore at bifurcation of intestine; testicles two, may be slightly lobate, near acetabulum; ovary posterior of and smaller than testicle, but anterior of transverse vitello-duct; ovary and testicles anterior of mass of uterine coils which extend to posterior end of the body.
Dicrocoeles (*Dicrocoelium*), 7.

Fascioles (Fasciola).

(6) Parasitic in liver or lungs of cattle. Body (figs. 28–30) flesh-colored, very large and thick, 20 to 100 mm. long by 11 to 26 mm. broad; anterior conical portion not very distinct from posterior portion; posterior extremity bluntly rounded; vitellogene glands situated ventrally of intestine; oesophagus generally one and one-half times as long as pharynx; eggs 109 to 168 μ by 75 to 96 μ.
The Large American Fluke (*F. magna*), p. 49.
Parasitic in liver or lungs of cattle, sheep, hogs, etc. Body (figs. 2 and 3) 18 to 51 mm. long (occasionally longer) by 4 to 13 mm. broad; anterior conical portion generally very distinctly bounded from posterior portion; posterior extremity bluntly pointed; vitellogene glands both dorsal and ventral of intestine; oesophagus rarely one and one-half times as long as the pharynx; egg 105 to 145 μ by 63 to 90 μ...... The Common Liver Fluke (*F. hepatica*), p. 29.

Parasitic in liver of Senegal cattle. Body (figs. 23 and 24) 26 to 38 mm. long by 6 to 8 mm. broad, flat, linguiform, sides of posterior portion nearly parallel for some distance but tapering toward posterior extremity; ventral sucker large and prominent; egg 143 to 151 μ by 82 to 88 μ.
The Narrow Liver Fluke (*F. hepatica angusta*), p. 48.

Parasitic in the Indian buffalo (Bos bubalis) *and cattle* (Bos taurus). *Body (figs. 25 and 26) 25 to 31 mm. long by 6 to 8 mm. broad; sides of body nearly parallel for some distance; posterior extremity somewhat rounded.*
The Egyptian Liver Fluke (*F. hepatica aegyptiaca*), p. 48.

Parasitic in the liver of giraffes and cattle (?). Body (fig. 27) 75 mm. long by 3 to 12 mm. broad, flat, oblong, lanceolate; anterior extremity cylindrical, attenuate; posterior extremity obtuse; sides nearly parallel for greater part of length; oral sucker 1.12 mm. in diameter, ventral sucker somewhat larger; oesophagus extends nearly to acetabulum; 8 to 10 lateral branches to each intestinal caecum; other organs agree with F. hepatica .... The Giant Fluke (*F. gigantica*), p. 49.

Dicrocoeles (Dicrocoelium).

(7) *Parasitic in liver of cattle, sheep, and swine. Body (figs. 36 and 37) lancet form, 4 to 10 mm. long by 1 to 2.5 mm. broad; anterior end much more attenuate than posterior end; semitransparent, spotted brown by eggs; cuticle without spines; oral sucker 0.5 mm. in diameter, subterminal; ventral sucker 0.6 mm. in diameter, one-fifth the length of body back of mouth; mouth followed by an oesophagus which, about halfway between oral sucker and acetabulum, immediately in front of cirrus pouch, branches into two simple intestinal caeca; the latter extend one each side to about the posterior quarter of the body; cirrus pouch present; cirrus long, filiform, straight; testicles lobed, one posterior to the other, and situated immediately posterior of acetabulum; uterus sinuous, very long, extending backward beyond the end of the intestine to posterior portion of body, then running forward in loops to genital pore, and rendered prominent by presence of brown eggs; vitellogene glands in marginal portion of middle third of body; eggs (fig. 38) 40 to 45 μ by 30 μ, containing embryo at time of oripont* .... The Lancet Fluke (*D. lanceatum*), p. 55.

Parasitic in pancreas of "cattle," Indian buffalo, and sheep, in Asia. Body (fig. 40) somewhat similar to the common fluke but proportionally broader and more pointed at distal extremity; 8 to 15 mm. long by 5 mm. broad; blood red in color; cuticle without spines; oral sucker subterminal; ventral sucker slightly larger than oral sucker, one-third the length of the body back of the mouth; pharynx, oesophagus, and intestines about the same as in D. lanceatum; cirrus-pouch pyriform; testicles irregularly lobed, in lateral portion of median field, on same transverse plane, near acetabulum; uterus of similar type to that of D. lanceatum; vitellogene glands only about one-fifth as long as body, situated in marginal portion of middle third; eggs ovoid, thick shelled, 44 μ to 49 μ by 23 μ to 30 μ.
The Pancreatic Fluke (*D. pancreaticum*), p. 57.

Blood Flukes (Family Schistosominae; *Genus* Schistosoma).

[Acetabulum penduculate; intestinal caeca unite or anastomose distal of acetabulum; male shorter, thicker, and broader than female, the margins curling ventrally to form canal for filiform female; testicular complex consists of a double series of four or more sacular bodies.]

(8) *Parasitic (fig. 41) in blood of man and cattle (?).* Male 4 to 14 mm. long by 1 mm. broad; female attains 13 to 20 mm. long by 0.28 mm. broad by 0.21 mm. thick; eggs ovoid to fusiform 120 to 197 μ long by 40 to 73 μ broad.
The Human Blood Fluke (*S. haematobium*), p. 58.

Parasitic (fig. 45) in blood of cattle. Body thicker than the Human Blood Flukes; the dorsal surface of the inner fold of the male is provided with a longitudinal groove (fig. 46) into which the end of the outer fold extends; eggs fusiform, 160 to 180 μ by 40 to 50 μ .... The Bovine Blood Fluke (*S. bovis*), p. 60.

Amphistomes (Amphistomidae).

(9) Parasitic in intestinal tract or gall ducts of sheep, cattle, zebu, and gayal. Pharynx without lateral sacs.... .... 10.

Parasitic in intestinal tract of zebu and gayal. Pharynx with two lateral sacs; the greater part of the ventral surface of posterior portion of body is covered with numerous papillae .... *Homalogaster*, 13.

(10) Parasitic in intestinal tract or gall ducts of sheep, cattle, and zebu. Ventral pouch very small or absent .... True Amphistomes (*Amphistoma*), 11.

Parasitic in intestinal tract of cattle, zebu, gayal, and Indian buffalo. Ventral pouch large, extending to posterior portion of body.

Pouched Amphistomes (*Gastrothylax*), 12.

True Amphistomes (Amphistoma).

(11) Parasitic in rumen of sheep and cattle. Body (figs. 49 and 50) 4 to 13 mm. long by 1 to 3 mm. broad; conical, pinkish white.. The Conical Fluke (*A. cervi*), p. 64.

Parasitic in gall bladder and hepatic ducts of zebu. Body lanceolate, 8 to 10 mm. long by 3 to 4 mm. broad, rather similar to the Conical Fluke, but somewhat flattened dorso-ventrally. .... *A. explanatum*, p. 67.

Parasitic in stomach of zebu. Body (fig. 56) somewhat similar to A. cervi, but more oral, and somewhat flattened dorso-ventrally; about 11 mm. long by 6.6 mm. broad.

A. bothriophorum, p. 67.

Pouched Amphistomes (Gastrothylax).

(12) *Parasitic in stomach of zebu and cattle. Body (figs. 57–62) reddish brown to grayish green, 9 to 15 mm. long by 4 to 5 mm. broad* .... *G. crumenifer*, p.67.

Parasitic in stomach of gayal. Body (fig. 63) pyriform, 10 mm. long by 5 mm. broad at posterior extremity .... *G. Cobboldii*, p. 67.

Parasitic in stomach of gayal and zebu. Body attains (fig. 64) 20 mm. long by 4 mm. broad; intestine only half as long as body .... *G. elongatum*, p. 67.

Parasitic in stomach of Indian buffalo. Body (figs. 65 and 66) deep red, cylindrical to conical, attains 7 to 10 mm. long by 2 to 2.5 mm. broad .... *G. gregarius*, p. 67.

Homalogaster.

(13) *Parasitic in caecum of gayal. Body (fig. 67) lanceolate; testicles small.*

     •   *H. paloniae*, p. 67.

Parasitic in caecum of "cattle" (= (?) zebu). Body — mm. long; oral sucker with digitate papillae; testicles lateral and divided into two equal lobes with irregular contours, so that there appear to be four testicular masses ...... *H. Poirieri*, p.67.

TAPEWORMS.
(Cestoda.)

(14) Larval tapeworms or bladder worms (figs. 68, 76, 84, 97, and 105) parasitic in muscles, liver, lungs, etc., but not in lumen of intestinal tract; the head, which is generally provided with hooks, lies inside the cyst; body unsegmented, generally surrounded by a cyst of connective tissue; no genital organs developed; these forms become adult in man and carnivorous animals, and are of great importance from the standpoint of meat inspection... Taeniinae, 15.

Adult tapeworms (fig. 111), found in the intestine of cattle, sheep, and swine (?), or in gall ducts of sheep .... Anoplocephalinae, 18.

Bladder Worms, or Larval Tapeworms (Hard-shelled Tapeworms, Subfamily Taeniinae; *Genus* Taenia).

(15) Parasitic in muscles, abdominal cavity, lungs, and liver. One head in each cyst ....: .... Bladder Worms (*Cysticercus*), 16.

Parasitic in any organ of body except intestine, most frequent in liver, lungs, and brain. Numerous heads may be present .... 17.

Bladder Worms (Cysticercus).

(16) Parasitic in cattle; found in the muscles, especially those of mastication, more rarely in lungs or liver. Body (fig. 68) spherical to elliptical, 2.5 to 10 mm. long by 3 mm. broad; whitish to gray, with a small yellowish spot due to the invaginated head; no hooks present; the bladder contains but little liquid. Transmissible to man.......... Beef Measle Bladder Worm (¹*C. bovis*), p. 71. Parasitic in swine; found in the muscles. Body (figs. 75 and 76) ellipsoid, 6 to 10 mm. long by 5 to 10 mm. broad, with a white spot corresponding to the invaginated head; head armed with a double row of 24 to 32 hooks of two different sizes (see description of adult, p. 84); the bladder contains but little liquid. Transmissible to man..... Pork Measle Bladder Worm (¹ *C. cellulosae*), p. 89. Parasitic in cattle, sheep, and swine; young stages in the liver, older stages found hanging into the body cavities, attached to omentum, etc. Bladder (figs. 84, 91, and 92) large, varying from size of a pea to that of a man's fist, occasionally attaining 160 mm. by 60 to 70 mm.; neck long; invaginated head armed with a double row of 28 to 44 (generally 36 to 38) hooks, of two sizes (see description of adult, p. 101); the bladder contains considerable liquid. Transmissible to dogs, but not to man.

The Thin Necked Bladder Worm (² *C. tenuicollis*), p. 96.

Coenurus and Echinococcus.

(17) *Parasitic in nervous system, especially the brain, of sheep and calves. Bladder (figs. 94 and 97) varies from size of a pea to that of a hen's egg, and is composed of a hydatid cyst (cuticle thin) which forms numerous small invaginations (as many as 500 in large specimens), in each of which a head derelops without the formation of brood capsules; head armed with a double row of 22 to 32 hooks, of two sizes (see description of, adult, p. 101). Transmissible to dogs, but not to man.*

The Gid Bladder Worm (¹ *Coenurus cerebralis*), p. 108.

Parasitic in any organ, particularly the liver and lungs of man, cattle, sheep, swine, etc. Bladder (figs. 101 and 105) varies from size of a pea to that of a child's head, assuming different forms, as described on p. 102; the hydatid cyst has a thick laminated cuticle; the heads are armed with a double row of 28 to 50 hooks, of two sizes (see characters of adult, p. 101), and develop in brood capsules, which are attached to the cyst wall. The adults develop in dogs, but not in man. This is the most important parasite of meat inspection.......... The Echinococcus Hydatid (² *Echinococcus polymorphus*), p. 113.

Adult tapeworms of Cattle, Sheep, and Swine (?) (Subfamily ³ Anoplocephalinae).

(18) Posterior border of segments not fringed; parasitic in the intestine....... 19. Posterior border of segments fringed (figs. 122 and 124). Parasitic in intestine and bile ducts of sheep. Genital pores double; strobila 15 to 30 cm. long; head large, 1.5 mm. broad, nearly square on apex view; neck flat, broad, and short; broadest segments measure 5 to 8 mm. wide by 0.4 to 0.6 mm. long, and are situated about 2 cm. from posterior end, the end segment showing a decided tendency to become longer and narrower; gravid segments attain 2.2 mm. in thickness; uterus single, transverse, but undulate with cornucopia-like egg pouches; testicles form a band in distal portion of median field; horns of pyriform body around embryo not developed.

The Fringed Tapeworm (*Thysanosoma actinioides*), p. 128.

¹ For characters of the adult form in man, see key, p. 84.
² For characters of the adult form in dogs, see key, p. 101.
³ This key to the adult forms is extremely artificial, as characters have been selected which will most easily enable a determination of the worms. For a key expressing more closely the true relations of the forms to each other, see Stiles & Hassall, 1893, p. 88, and Stiles, 1896, p. 214.

(19) Genital pores double, segments generally quite broad; pyriform body well
developed.   Parasitic in cattle and sheep...................... *Moniezia*, 20.
*Genital pores usually single, rarely double, and then only in strobilae, which contain
single-pored segments in the majority.   Parasitic in sheep, cattle (?), and swine (?).
Strobila 1 to 2 meters long; head 0.5 to 1 mm. broad; neck absent or present; most
of the segments broader than long, only the posterior segments longer than broad;
unripe segments often present a zigzag appearance; largest segments 5 to 6.5 mm.
broad by 1 to 2 mm. long; testicles divided into two groups and confined for the
most part to the lateral fields; uterus same as in the Fringed Tapeworm.*
                                    Giard's Thysanosoma (*Th. Giardi*), p. 129.
Genital pores single; segments very narrow.   Parasitic in sheep and cattle (?).
                                •   *Stilesia*, 24.
                    Moniezia.

(20) *Interproglottidal glands absent*.................................... Alba group, 21.
Interproglottidal glands linear (fig. 113).................. Planissima group, 22.
Interproglottidal glands circular, groups around blind sacs (fig 118).
                                                          Expansa group, 23.
*Neck absent; head large, decidedly lobed; strobila 40 cm. long; segments attain 8 mm.
in breadth.   Doubtful species, parasitic in sheep*.......... *Moniezia nullicollis.*
(21) *Parasitic in sheep.   Strobila nearly half a meter long by 2.5 mm. broad; gravid segments may attain 2.5 mm. broad by 5 mm. long; genital pores in middle or anterior
half of lateral margin; cirrus pouch about 0.18 mm. long; eggs 60 μ, pyriform body
20 μ, horns end in a knob*.............-..... Vogt's Moniezia (*M. Vogti*), p. 127.
*Parasitic in sheep and cattle.   Strobila 0.60 to 2.5 meters long; head subquadrangu-
lar, 1.15 to 1.4 mm. broad; neck 1.5 to 5.3 mm. long; gravid segments attain 8 to 14
mm. broad by 2 to 6.5 mm. long by 1.5 mm. thick; testicles arranged in a quadrangle;
eggs 60 to 88 μ, bulb of pyriform body 16 to 24 μ, horns 8 to 30 μ.*
                                    The White Moniezia (*M. alba*), p. 127.
(22) Parasitic in cattle and sheep.   Strobila (fig. 111) 1 to 2 meters long; yellowish;
head 0.4 to 0.9 mm. broad; neck thin, short or long; segments always broader
than long; gravid segments attain 12 to 26 mm. broad by 1 to 1.75 mm. long,
generally thin and flat; interproglottidal glands large and very distinct; testicles arranged at first in two triangles, in older segments in a quadrangle;
400 to 600 testicles present in a segment; eggs 63 μ, bulb of pyriform body 20 μ,
horns 24 μ...... .... The Flat Moniezia (*M. planissima*), p. 127.
*Parasitic in sheep and cattle.   Strobila attains 4 meters in length; head about 1 mm.;
neck 2 to 2.5 mm. long; suckers very distinctly lobed and sharply separated from
neck; segments always broader than long; gravid segments may attain 12 mm. broad
by 3 mm. long by 2 mm. thick; interproglottidal glands extremely indistinct; eggs
80 to 85 μ, pyriform body 18 μ..* Van Beneden's Moniezia (*M. Benedeni*), p. 128.
*Parasitic in sheep.   Strobila 1.5 to 2 feet long; head square, 0.9 mm.; gravid seg-
ments attain 8 mm. broad by 1.5 mm. long; but the end segments may measure 6
mm. broad by 2 mm. long; testicles arranged in a quadrangle; interproglottidal
glands small; eggs 55 to 65 μ*..... Neumann's Moniezia (*M. Neumanni*), p. 128.
(23) Parasitic in cattle and sheep.   Strobila (fig. 116) attains 4 to 5 meters in length:
anterior portion usually whitish, posterior portion usually yellowish; head
0.36 to 0.7 mm. broad; segments always much broader than long, gravid segments attaining 16 mm. in width and are quite thick; end segments never as
long as broad; testicles usually arranged in a quadrangle, rarely in two triangles except in younger segments; eggs 50 to 60 μ, bulb of pyriform body
20 μ................................. The Broad Moniezia (*M. expansa*), p. 128.
*Parasitic in sheep.   Strobila (fig. 120) attains 1.6 to 2 meters in length; cream
to whitish in color; head 0.6 to 0.7 mm. broad; neck filiform, 2 mm. long; segments generally broader than long, rarely over 6 mm. broad by 2 mm. long;
although end segments are occasionally found which are square or even slightly*

longer than broad; testicles usually arranged in two triangles; eggs 52 to 60 μ, bulb of pyriform body 20 to 24 μ, horns 12 to 15 μ.

The Triangle Moniezia (*M. trigonophora*), p. 128.

Stilesia.

(24) *Strobila transparent, whitish or grayish yellow, 45 to 60 mm. long, not over 2.5 mm. broad; head 0.5 to 1 mm. broad; median portion of median field transparent; two lateral cornucopia-like egg pouches present in each segment.*

The Globipunctate Tapeworm (*S. globipunctata*), p. 130.

Strobila attains nearly 3 meters in length, but not over 3 mm. in breadth; head 1.5 to 2 mm. broad; median field occupied by transverse uterus.

The Centripunctate Tapeworm (*S. centripunctata*), p. 130.

FLUKES, OR TREMATODES (Order Trematoda).

The following technical description shows the systematic position and general structure of the flukes under discussion:

[Suborder Malacocotylea: Digenea.   Families Fasciolidae and Amphistomidae. See figs. 3, 29, 30, 37, 41, 42, 43, and 50.]

With the exception of the Blood Flukes (*Schistosoma*), they are all hermaphrodites. They are flat or conical worms, always longer than broad; on the anterior extremity is situated the mouth, surrounded by a muscular organ, known as the *oral sucker* and curved slightly ventrad. There is a second sucker (the *acetabulum*), which is situated in the median ventral line; in the Fasciolidae the acetabulum is generally found on the anterior half of the body, while in the family Amphistomidae it is at or near the posterior extremity. The surface of the worms is generally more or less covered with minute spines, or tubercles.

The *digestive tract* consists of the mouth, a short oesophagus, and two blind sacs (*intestinal caeca*), which represent the true intestine. The anterior portion of the oesophagus is generally connected with the mouth by a muscular bulb (the *pharynx*); the posterior extremity bifurcates, one branch being connected with each intestinal caecum. The intestinal sacs are usually simple elongated tubes, but in the genus *Fasciola* they branch freely (fig. 29). In *Schistosoma* the two caeca unite after passing the acetabulum. An anus is never present.

Genital organs.—The *genital pore* is in the ventral median line in all species here described, the male copulatory organ (*cirrus* or penis) lying very close to the female opening (*vulva*). *Male organs:* A cirrus is frequently seen extruded from the genital pore, and in those cases it appears as a curved organ, varying in size according to the species; usually the cirrus is invaginated in the *cirrus pouch*. Through its center runs a canal (the *ductus ejaculatorius*) which receives the spermatozoa from a *vesicula seminalis*. The latter is partially or entirely included in the pouch; at its posterior end it receives the two *vasa deferentia*, through which the spermatozoa are conducted from the testicles. The testicles, generally two in number, one right and one left, are more or less round, lobed, or branched. *Female organs:* The vulva leads into a canal, the anterior portion of which is known as the *metraterm;* this is continued as the *uterus*, which forms more or less numerous folds in the median portion of the body and finally leads to the so-called *shell-gland* which may frequently be seen in fresh specimens (*F. magna* and others) as a round body a short distance posterior of the acetabulum. In the center of the shell gland is a canal (the *ootyp*), in which four canals (*uterus, oviduct, Laurer's canal,* and *vitello-duct*) come together. The *ovary* in some species is globular, in others branched, and connects with the ootyp through the oviduct. The *Laurer's canal* runs from the ootyp dorsad in curves and opens to the exterior on the dorsal surface; its function is still doubtful, but homologically it represents the uterus of cestodes. The *vitellogene glands* are two in number and are situated laterally of the longitudinal intestinal tubes; they vary in size in different species, are generally quite elongated, and are composed of numerous branches much like a bunch of grapes in form, all of which connect with a longitudinal *vitello-duct* (one on each side of the body); these longitudinal ducts are in turn connected by a pair of transverse ducts which unite in the median line, immediately posterior of the shell-gland, to form a common reservoir; this in turn empties into the ootyp through the short vitello-duct mentioned above. The vitellogene glands produce yolk cells which are associated with the true ovum to form the eggs.

Excretory system.—At or near the posterior extremity, generally somewhat dorsally, is situated a small pore (*porus excretorius*), which leads into a *median terminal vesicle;* this latter gives off longitudinal branches; these in turn give off secondary branches which ramify throughout the body, each small branch ending in an excretory organ.

Nervous system.—A set of ganglia is found at each side of the pharynx; these ganglia are connected by a dorsal commissure and give off numerous nerves to various parts of the body. The largest nerves are the two ventral longitudinal nerves which run antero-posteriorly, and can frequently be seen in fresh specimens.

Development.—See p. 30.

Cattle (*Bos taurus*) are alleged to be infested with fifteen kinds of flukes, only two of which, the Large American Fluke and the Common Liver Fluke, are positively known to occur in the United States. Osler has found the Conical Fluke at Montreal, where it was not uncommon; he also found the same parasite in cattle in Nova Scotia.

Sheep (*Ovis aries*) are infested with five known species of flukes, only one of which, the Common Liver Fluke, is known to be in the United States; the Conical Fluke, as stated above, is found in Canada.

Hogs (*Sus scrofa domestica*) harbor three known species of flukes, only one of which, the Common Liver Fluke, is found in the United States. Willach (1893) has described a *Monostomum hepaticum suis* from the liver of hogs; this supposed fluke is evidently a partially developed bladder worm (*Cysticercus tenuicollis*) (see p. 96).

DISTOMES (Flukes of the Family Fasciolidae).

Hermaphroditic Distomes (Flukes of the Subfamily *Fasciolinae*).

AGAMIC, OR IMMATURE, DISTOMES (Genus Agamodistomum).

This is a purely artificial group, of biologic rather than systematic nature. One immature fluke is occasionally found encysted in the muscles of hogs.

1. The Muscle Fluke of Swine (*Agamodistomum suis*).

[Fig. 1.]

SYNONYMY.—*Distomum musculorum suis* Duncker, 1896.
BIBLIOGRAPHY.—Duncker (1896).

This small (0.5 mm. to 0.7 mm. long by 0.2 mm. broad) parasite was discovered in 1881 by G. Leunis (a trichina inspector in Saxony), and has since been found by several other trichina inspectors of Germany. As it appears never to have been binomially named, I propose to call it

Agamodistomum suis. The worm lies free or encysted in the connective tissue between the muscle fibers; it is exceedingly rare and is of no known practical importance in meat inspection, except that in a superficial and careless microscopic examination it might be mistaken for sarcosporidia, or possibly for trichinae. Nothing is known of its life history, but it is supposed to be a purely accidental parasite in swine. We are not aware of its ever having been recorded in this country.

FASCIOLES (Distomes of the Genus Fasciola).

The genus *Fasciola* contains the large, flounder-like, parasites found especially in the liver of herbivorous animals and known under the general term *"liver flukes."* Of these Fascioles, or *"liver flukes,"* we find two forms in American cattle (*F. magna* and *F. hepatica*), one form (*F. hepatica*) in American sheep, while a third form (*F. Jacksoni*[1]) has been found in North America, South America, and in India in the liver of elephants, and a fourth form (*F. gigantica*[1]) is described by Cobbold from the liver of the giraffe. It is quite generally admitted that these Fascioles, owing to their larger size, are more harmful than other flukes.

Until a short time ago it was supposed that we had but one form of fluke in American cattle, but Hassall (1891) and Francis (1891) showed, almost simultaneously, that two distinct forms are found, one form (*F. hepatica*) being present in the liver, very rarely in the lungs, the other (*F. magna*), a much larger worm, infesting both liver and lungs.

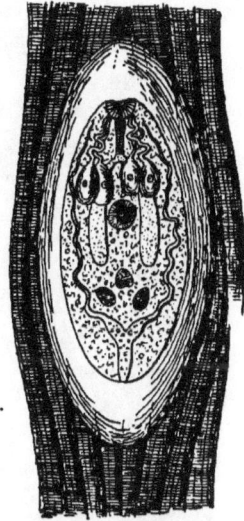

Fig. 1.—The Muscle Fluke (*Agamodistomum suis*), occasionally found in the muscle of swine. (After Leuckart, 1889, p. 155, fig. 86.)

2. The Common Liver Fluke (*Fasciola hepatica*) of Cattle, Sheep, Swine, etc.

[Figs. 2-22.]

For anatomical characters, compare fig. 3 with key, p. 21.

Fig. 2.—The Common Liver Fluke (*Fasciola hepatica*), natural size (original).

VERNACULAR NAMES.—English, *Common Liver Fluke;* German, *Leberegel, Leberwurm, Schafegel;* Dutch, *Botten, Leverworm;* Danish, *Faareflynder;* Swedish, *Levermask;* French, *Doure hépatique, fasciole;* Italian, *Biscuola, distoma epatico;* Spanish, *Caracolillo.*

SYNONYMY.—*Fasciola hepatica* Linnaeus, 1758; *Planaria latiuscula* Goeze, 1782; *Distoma hepaticum* (Linnaeus) Abildgaard (?); *Fasciola humana* Gmelin, 1790; *Distoma (Cladocoelium) hepaticum* (Linnaeus) of Dujardin, 1845; *Fasciolaria hepatica* (Linnaeus) anonymous, 1845; *Distomum hepaticum* (Linnaeus) Diesing, 1850; *Distomum (Fasciola) hepaticum* Linnaeus of Leuckart, 1863; *Cladocoelium hepaticum* (Linnaeus) Stossich, 1892.

BIBLIOGRAPHY.—For bibliography, see Hassall (1894) and Huber (1894). For more technical discussion of species, see Leuckart (1889, pp. 179-328).

GEOGRAPHICAL DISTRIBUTION.—Cosmopolitan.

HOSTS.—Man, cattle, sheep, swine, and other animals. (See pp. 137-143.)

[1] For a discussion of these forms, see Stiles, 1894-1895.

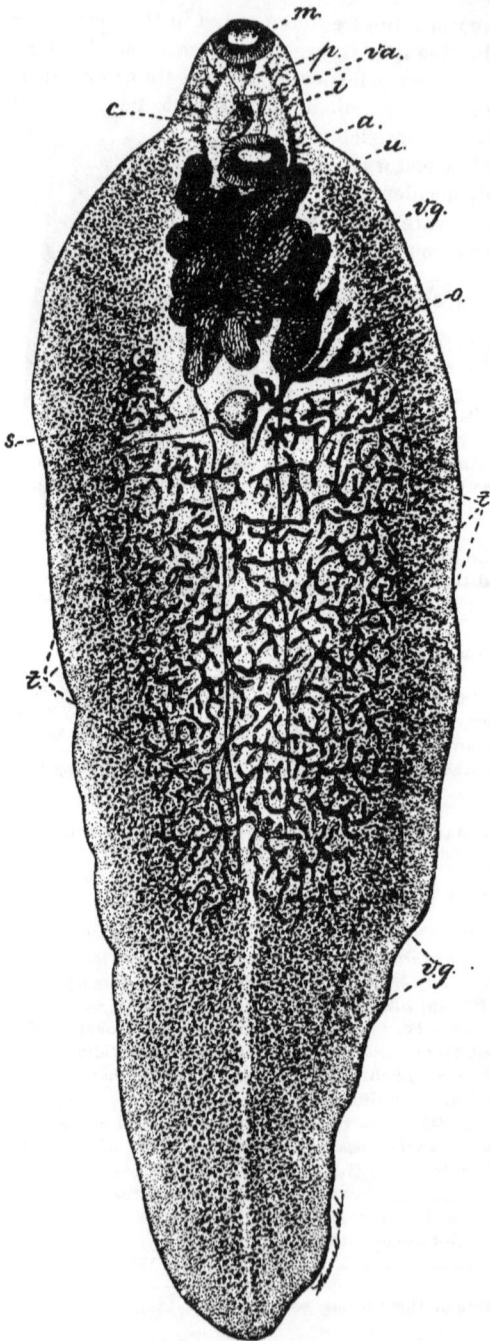

FIG. 3.—The Common Liver Fluke (*Fasciola hepatica*), enlarged to show the anatomical characters: *a*, acetabulum; *c*, cirrus pouch; *i*, intestine; *m*, mouth with oral sucker; *o*, ovary; *p*, pharyngeal bulb; *s*, shell gland; *t*, profusely branched testicles; *u*, uterus; *va*, vagina; *vg*, profusely branched vitellogene gland. (After Stiles, 1894, p. 300.)

Life history.—The life cycle of this fluke, as determined by the investigations of Creplin (1837), Weinland, Leuckart (1863, 1879 1880, 1881, 1882), and Thomas (1882, 1883), is exceedingly interesting; at the same time it is very complicated, for the adult parasite, instead of producing young similar to itself and capable of developing directly into adults in cattle, produces eggs which develop into organisms totally different from the adult form living a parasitic life in other animals. In scientific language, the parasite is subject to an alternation of generations, together with a change of hosts. The following summary of the life history will make this point clear:

(a) *The adult hermaphroditic worm* (figs. 2 and 3), the characters of which are given on p. 22, fertilizes itself (although a cross fertilization of two individuals is not impossible) in the biliary passages of the liver, and produces a large number (estimated at 37,000 to 45,000) of eggs.

(b) *Eggs* (figs. 4 and 5).—Each egg is composed of the following parts: (1) A true germ cell, which originates in the ovary and is destined to give rise to the future embryo; (2) a number of vitelline or yolk cells, which are formed in a specialized and independent portion (vitellogene gland) of the female glands; instead of developing into embryos, the yolk cells form a follicle-like covering for the true germ cell and play an important rôle in the nutrition of the latter as it undergoes further development; (3) a shell surrounding the germ cell and vitelline cells, and provided at one end with a cap or operculum. The eggs escape from the uterus of the adult through the vulva, are carried to the intestine of the host with the bile, then pass through the intestines with the contents of the latter, and are expelled from the host with the faecal

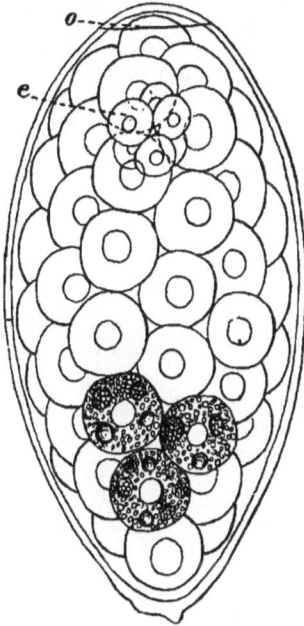

FIG. 4.—Egg of the Common Liver Fluke (*Fasciola hepatica*) examined shortly after it was taken from the liver of a sheep; at one end is seen the lid or operculum, *o;* near it is the segmenting ovum, *e;* the rest of the space is occupied by yolk cells which serve as food; all are granular, but only three are thus drawn. × 680. (After Thomas, 1883, p. 281, fig. 1.)

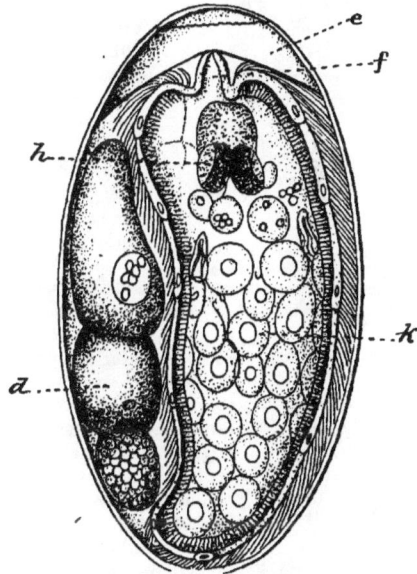

FIG. 5.—Egg of the Common Liver Fluke containing a ciliated embryo (miracidium) ready to hatch out: *d*, remains of food; *e*, cushion of jelly-like substance; *f*, boring papilla; *h*, eye-spots; *k*, germinal cells. × 680. (After Thomas, 1883, p. 283, fig. 2.)

matter. Many of them become dried and then undergo no further development, but others are naturally dropped in the water in marshes, or, being dropped on dry ground, they are washed into the water by the rain, or are carried to a more favorable position by the feet of animals pasturing or passing through the fields. After a longer or shorter period of incubation, which varies with the temperature, a ciliated embryo (*miracidium*) is developed. At a temperature of 20° to 26° C. the miracidium may be formed in 10 days to 3 weeks; at a temperature of 16° C. the development takes 2 to 3 months; at 38° C. it ceases entirely. Experiments have shown that as long as these eggs remain in the dark the miracidium will not escape from the eggshell; accordingly it will not escape during the night. When exposed to the light, however, or when suddenly brought into contact with cold water, the organism bursts the cap from the eggshell, crawls through the opening, and becomes a—

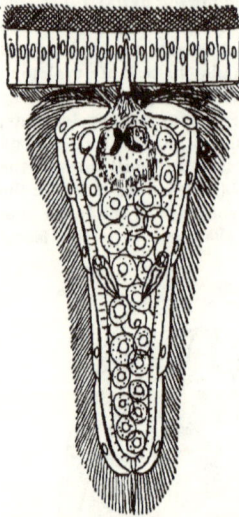

FIG. 6.—Embryo of the Common Liver Fluke (*Fasciola hepatica*) boring into a snail. × 370. (After Thomas, 1883, p. 285, fig. 4.)

(c) *Free-swimming ciliated miracidium* (fig. 6).—As already stated, this organism is entirely different from its mother. It measures about 0.15 mm. long; it is somewhat broader in its anterior portion than in its posterior portion; on its anterior extremity we find a small eminence known as a boring papilla; the exterior surface of the young worm is covered with numerous cilia, which by their motion propel the animal through the water; inside the body we find in the anterior portion a simple vestigial intestine and a double ganglionic mass, provided with a peculiar pigmented double cup-shaped eye-spot; in the posterior portion of the body cavity are found a number of germ cells, which develop into individuals of the next generation.

Swimming around in the water, the miracidium seeks out certain snails (*Limnaea truncatula, L. oahuensis, L. rubella*, see p. 43), which it immediately attacks (fig. 6). The miracidium elongates its papilla and fastens itself to the feelers, head, foot, or other exterior soft portion of the body of the snail; some of the parasites enter the pallial (lung) cavity and attach themselves there. After becoming securely fastened to the snail the miracidium discards its ciliated covering and shortens to about half its former length (0.07 mm. to 0.08 mm.). The parasites now bore their way into the body of the snail and come to rest in the liver, or near the roof of the pallial cavity, etc., the movements

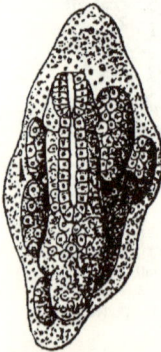

FIG. 7.—Sporocyst of the Common Liver Fluke which has developed from the embryo, and contains germinal cells. × 200. (After Leuckart, 1889, p. 109, fig. 67 B.)

gradually cease, and we have before us the stage known as the—

(d) *Sporocyst* (figs. 7 and 8).—The eye-spots, ganglionic swellings, and vestigial intestine become more and more indistinct and are finally lost. The sporocyst grows slowly at first, then more rapidly, and at the end of 14 days or so measures about 0.5 mm. The germ cells mentioned as existing in the posterior portion of the miracidium now develop into individuals of a third generation, known as—

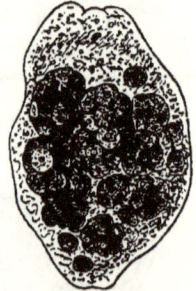

FIG. 8.—Sporocyst of the Common Liver Fluke, somewhat older than that of fig. 7, in which the germinal cells are giving rise to rediae. × 200. (After Leuckart, 1889, p. 109, fig. 67 C.)

(e) *Rediae* (figs. 9 and 10).—The rediae escape from the sporocyst when the latter are from two weeks (in summer) to four weeks (in late fall) old. Upon leaving the body of the sporocyst they wander to the liver of the snail, where they grow to about 2 mm. long by 0.25 mm. broad. Each redia consists of a cephalic portion, which is extremely motile, and which is separated from the rest of the young worm by a ridge; under the latter is situated an opening, through which the next generation (cercariae) escape. The posterior portion of the worm is provided, at about the border of the third and the last fourths of the body, with two projections. There is a mouth with pharynx situated at the anterior extremity, the pharynx leading into a simple blind intestinal sac. The redia, as well as the sporocyst, may be looked upon as a female organism, and in its body cavity are found a number of germ cells, which develop into the individuals of the next generation, known as—

(*f*) *Cercariae* (figs. 11-13).—These organisms are quite similar to the adult parasites into which they later develop. The body is flat, more or less oval, and provided with a tail inserted at the posterior extremity. The oral sucker and acetabulum are present as in the adult, but the intestinal tract is very simple; on the sides of the body are seen two large glands, but the complicated genital organs of the adult are not visible. The cercaria leaves the redia through the birth opening, remains in the snail for a longer or shorter time, or passes out of the body of the snail and swims around in the water. After a time it attaches itself to a blade of grass (fig. 12) or some other object, and forms a cyst around itself with material from the large glands, at the same time losing its tail. It now remains quiet until swallowed by some animal. Then, upon arriving in the stomach—of a steer, for instance—the cyst is destroyed, and the young parasite wanders through the gall ducts or, as some believe, through the portal veins to the liver, where it develops into the adult hermaphrodite.

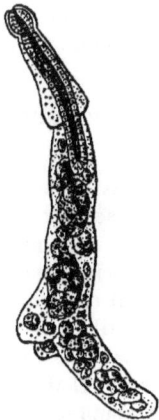

FIG. 9.—Redia of the Common Liver Fluke (*Fasciola hepatica*), containing germinal cells which are developing into cercariae. × 150. (After Leuckart, 1889, p. 209, fig. 129 *A*.)

From the above we see that this parasite runs through three generations, namely:

(1) O v u m, miracidium, and sporocyst....first generation.

(2) Redia...second generation.

(3) Cercaria and adult....third generation.

During this curious development, which lasts about 10 to 12 weeks, there is a constant potential increase in the number of individuals, for each sporocyst may give rise to several (5 to 8) rediae, each redia to a larger number (12 to 20) cercariae, and each adult to an enormous number (37,000 to 45,000) of eggs. This unusual fertility of the animal is necessary because of the complicated life history and the comparatively small chance any one egg has of completing the entire cycle.

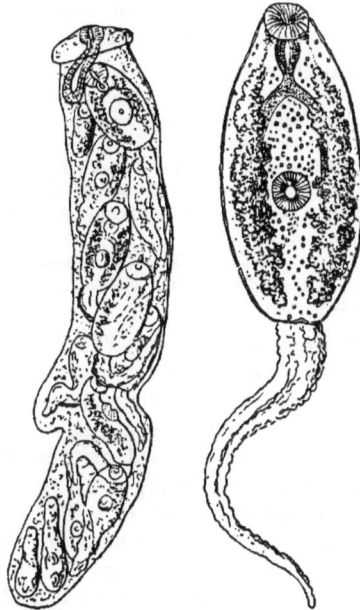

FIG. 10.—Redia of the Common Liver Fluke, with developed cercariae. × 150. (After Leuckart, 1889, p. 270, fig. 130.)

FIG. 11.—Free cercaria of the Common Liver Fluke, showing two suckers, intestine, large glands, and tail. (After Leuckart, 1889, p. 279, fig. 137.)

Hosts.—An interesting and, from an agricultural standpoint, an important matter connected with this fluke is that it is found in a large number (about 25) of domesticated and wild animals, and this fact probably explains to some degree the wide geographical distribution of the parasite.

THE EFFECTS OF THE COMMON LIVER FLUKE UPON CATTLE, SHEEP, AND SWINE.

This worm is one of the most important and dangerous parasites with which the stock raiser has to deal, since it produces a disease which often results in heavy loss of live stock, especially of sheep. Although it does not seem as yet to have caused any such serious epizootics in this country as have been reported in Europe, sweeping out or greatly retarding the live-stock industry, we should not wait until such an occasion arises before we consider the importance of this subject. We know that *F. hepatica* is present in the country; furthermore, that it is common in some places (Texas and elsewhere), and we would do well to inquire into the injury which other countries have sustained as a warning that we must not totally ignore its presence among us.

The following are among the most important outbreaks[1] recorded:

Wernicke (1886) records that not less than 1,000,000 sheep died of fluke disease in the southerly provinces of Buenos Ayres during 1882; in 1886 more than 100,000 head died in Tandil during eight months.

Youatt estimated the annual loss in Great Britain at 1,000,000 sheep. For 1879 and 1880, a loss of 3,000,000 head per year was estimated for England alone.

During 1876, Slavonia lost 40 per cent of her cattle from distomatosis.

In 1830, England lost 3,000,000 sheep from this disease, estimated at a value of $20,000,000.

In 1829 and 1830, 5,000 of the 25,000 cattle of Montmédy perished; in Verdun, 2,200 cattle and nearly 20,000 sheep, out of 20,000 cattle and 50,000 sheep, succumbed to the parasite.

Names of the disease.—The presence of these flukes in the liver of animals gives rise to a disease known under the various names of rot, liver-rot, rot-dropsy, fluke disease, aqueous cachexia, cachexia aquosa verminosa, fascioliasis, distomatosis, etc.

The term *rot*, as used by farmers and by some veterinarians, is an exceedingly broad one; in many parts of this country almost any disease of sheep is called rot. We have met nodular disease of the intestine and other diseases under this term. On this account it must not be supposed that every article on rot refers to liver-fluke disease.

Symptoms.—There is no one special symptom which is characteristic of this disease and absent from all others; in fact liver rot in its various stages might easily be mistaken for other parasitic complaints.

(A) *The disease in sheep.*—Gerlach has divided the malady into four periods, and although this division is more or less artificial, since the different stages gradate imperceptibly into each other and are obscured

---

[1] For a more complete list of epizootics, see Hassall, 1894.

on account of the constant liability to further infection, we give Gerlach's scheme here as a convenient diagram of the disease:

These symptoms are taken chiefly from sheep, but the same description applies in
a general manner to the same disease in other animals:

(1) *Period of immigration* (stage of traumatic hepatic inflammation, inflammatory
swelling of the liver).—July to September, lasting about 13 weeks. This is the
period of infection, but as the
symptoms are not generally
very pronounced (the pathological lesions produced by
the flukes not having as yet
affected the system of the
host) it generally escapes
notice. At first a redness of
the eyes, which, however,
soon disappears; paleness.
Death from apoplexy some-
times occurs. The presence
of the flukes in the liver irri-
tates this organ and causes an
increased blood supply (hy-
peraemia) and consequent
enlargement of the liver. The
surface is smooth, marked
with small openings, out of
which may be pressed a

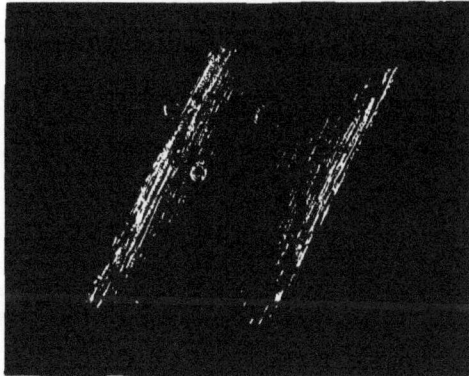

FIG. 12.—Portion of a grass stalk with three encapsuled cercariae
of the Common Liver Fluke (*Fasciola hepatica*). × 10. (After
Thomas, 1883, p. 291, fig. 13.)

bloody serum, and around these openings there is frequently an inflammation of the
peritoneum (localized peritonitis). Gall ducts still about normal; gall more or less
bloody; hemorrhagic centers in parenchyma; bloody serous exudate in abdominal
cavity, in which flukes are occasionally found. No eggs present as yet in droppings.

(2) *Period of anaemia.*—September to December, 6 to 12 weeks. The visible
mucous membranes (around the eyes, nose, and gums) and the skin are paler than
usual. Animals have a tendency to fatten. Appetite may be very good, but afterwards diminishes and rumination becomes irregular; slight oedema; bare skin soft
to the touch, loose and pasty; eyes become "fat," i. e., they
are partially closed, the conjunctiva becoming puffy; gradual
loss of strength; fever and accelerated respiration; death in
this stage seldom.

Liver pale, increased considerably in size, especially in
thickness; its capsule rough, opaque; its parenchyma soft
with an appearance like porphyry, with hemorrhagic
centers; here and there channels caused by parasites;
numerous eggs in faeces.

FIG. 13.—Isolated encysted
cercaria of the Common
Liver Fluke. × 150.
(After Leuckart, 1889, p.
286, fig. 142.)

(3) *Period of emaciation* (stage of atrophy of the liver).—
January to May. Disease is at its height; extreme anaemia
and emaciation; respiration feeble and quickened; temperature variable; abortions frequent; "puffiness" (oedema)
especially frequent under the jaws; mortality high.

Atrophy of liver in various stages; gall ducts greatly thickened, frequently with
calcareous incrustations; petechiae beneath endocardium; bile thick, dirty brown,
with numerous eggs.

(4) *Period of emigration of the flukes.*—May to July. The flukes leave the liver and
are passed with the droppings. The symptoms diminish, but the scars, the result of
the inflammatory processes, remain.

Zündel makes a slightly different division of the periods of the disease, but, as in the division proposed by Gerlach, it is not to be followed too rigidly, as the different periods are not sharply defined from one another. Zündel's four periods are:

First period.—Stage of inflammation, inflammatory swelling of the liver: August to October. The presence of the flukes causes an irritation; profuse flow of gall mixed with blood. Generally passes unnoticed.

Second period.—Stage of contraction of the liver: September to November, 6 to 12 weeks. Flukes collected in groups partially obstruct the bile ducts, whose irritated mucosa is thickened; anaemia, cachexia, general weakness, discoloration of the tissues.

Third period.—Stage of atrophy of the liver: January to March. Cachexia; high mortality. Flukes mature; gall ducts greatly thickened and hardened. Liver atrophies in some places, swells in others.

Fourth period.—Stage in which the flukes leave the liver: April to June.

(B) *The disease in cattle.*—The first symptoms are generally overlooked, the disease not attracting attention until the appetite is diminished, rumination becomes irregular, the animals become hidebound, and the coat dull and staring. The staring coat is due to the contraction of the muscles of the hair follicles. The visible mucous membranes become pale, eyes become dull, there is running at the eyes, and the animal gradually becomes emaciated. As the disease advances the milk supply is lessened, fever appears, there is generally great thirst, but the appetite almost ceases; oedematous swellings appear on the belly, breast, etc.; diarrhoea at first alternates with constipation, but finally becomes continuous. The disease lasts from 2 to 5 months, when the most extreme cases succumb.

Ostertag (1895, p. 357) states that most of the European cattle are infested with liver flukes, but that even when a large number are present the nourishment of the cattle is not disturbed. Thickening of the gall ducts, so that a so-called "Medusa's head" forms on the surface of the liver toward the stomach, appears in even well-nourished animals; even in cases of a cirrhosis of the liver it is seldom that any effect upon the cattle's health can be noticed, and as long as a portion of the liver tissue, about twice the size of the fist, remains intact the nourishment of the animal may be comparatively good. Ostertag, in all of his experience, has never seen a generalized oedema in slaughtered cattle as a result of fluke invasion, and even in the heaviest infections of young cattle he has noticed only emaciation.

(C) *The disease in hogs.*—The Common Liver Fluke is a comparatively rare parasite in swine and apparently of very little importance.

Pathology.—The pathological lesions are directly dependent upon the presence of the flukes in the body, and as the liver is the chief abode of the parasites, we should accordingly expect to find that this organ is more affected than any other, and the seat of the primary lesions; also that the symptoms and changes noticed in other organs are in nearly all cases directly dependent upon the changes in the liver; furthermore,

that the extent of the lesions is dependent upon the number of parasites present. The size of the worms and the size of the spines found
on them are two important factors in determining the extent of the
lesions.

By their presence and wanderings in the gall ducts the parasites irritate the
mucosa and cause an inflammation accompanied by an increased secretion, leading
to a desquamative catarrh; this inflammation causes a thickening of the mucosa
with growth of its glandular elements (glandular hyperplasia) and submucosa.
The young parasites make their way into the smaller ducts, rolling their body dorsally and here singly, or in the larger ducts in groups, they cause a dilatation of the
ducts, in some cases forming cysts. There is a hyperplasia of the connective tissue

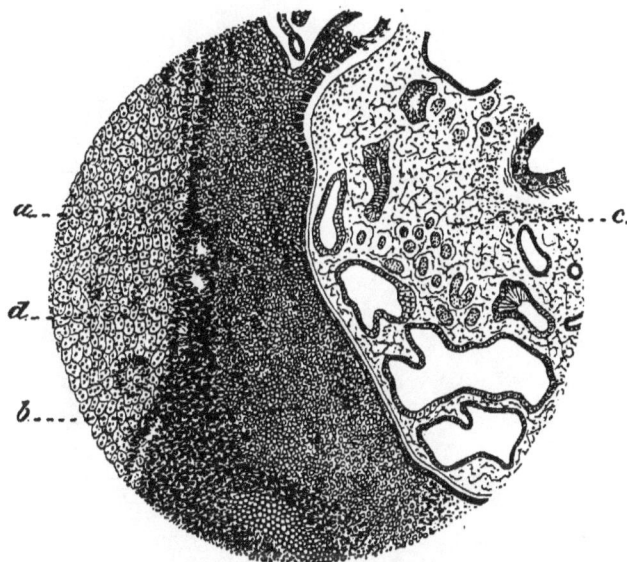

Fig. 14.—Drawing from a microscopic preparation showing a hemorrhage in the parenchyma of the
liver caused by the Common Liver Fluke (*Fasciola hepatica*); a, atrophic liver tissue; b, round cell
infiltration; c, a portion of the parasite; d, hemorrhage. (After Schaper, 1890, Pl. I, fig. 1.)

and a cellular infiltration, together with an increased development of the blood
capillaries. The inflammatory process extends from the duct walls to the interlobular
connective tissue, accompanied by atrophy of the parenchyma. A slight atrophy of
the parenchyma, with an extensive hypertrophy of the connective tissue and an
extensive infiltration and increased blood supply, naturally causes an increase in the
size of the liver. With the decrease of the hyperplastic tissue and the consequent
compression and destruction of the capillaries the cirrhotic and atrophic processes
become evident. An advancing hyperplasia of the connective tissue destroys the
parenchymatic cells of the lobules, leaving in many cases only a clump of gall pigment as evidence of a former lobule. Gradually a smaller or larger portion of the
liver is changed into a mass of cicatricial tissue surrounding stiff tubes—the meta-
morphosed gall ducts.

Besides the lesions thus far described, due for the most part to the changes in the
gall ducts, other changes are found due directly to the action of the parasites upon
the parenchyma of the liver, namely, a breaking down of the liver tissue, paren-

chymatic hemorrhages, pus infiltrations, and abscess formations. The flukes may break through a smaller gall duct, or may penetrate one of the larger ducts at a weak point, and wander directly through the soft glandular tissue; the mechanical injury to the tissue results in its necrosis; blood vessels are also injured, giving rise to multiple hemorrhages, which may discharge through the gall ducts and aid in producing the general anaemia. Inflammation naturally follows the flukes in their wanderings, leading to a liquefaction of the tissue and formation of abscesses, in which bacteria (streptococci and staphylococci) are found, the organisms having come from the inflamed bile ducts. With this inflammation going on it is but natural that the walls of some of the blood vessels should be affected, thus making it possible for the flukes to gain access to the circulation, with which they might be

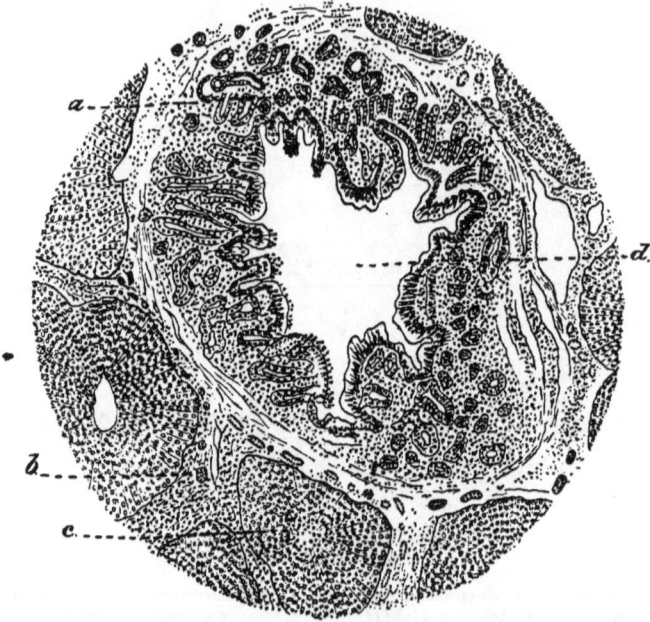

FIG. 15.—Drawing from a microscopic preparation showing the glandular hyperplasia of the mucosa of a gall duct caused by the Common Liver Fluke (*Fasciola hepatica*): *a*, hypertrophied submucosa; *b*, interstitial connective tissue; *c*, compressed lobule; *d*, lumen of the gall duct; thickened fibrous wall of the gall duct. (After Schaper, 1890, Pl. I, fig. 2.)

carried to various parts of the body, lungs, brain, etc., causing endophlebitis, endarteritis, ruptures, thrombosis, emboli, abscesses, etc.; pyaemic or septicopyaemic processes may extend from the liver, and finally the flukes in their wanderings may perforate the capsule of the liver, causing perihepatitis or peritonitis.

These various pathological lesions naturally act upon the circulatory system. The branches of the portal veins and vena cava are compressed or obliterated to a certain extent, and ascites and oedema follow.

The bile is greatly changed, becoming more or less thick, greenish brown, or dirty red, and containing epithelial, parenchymatic, and blood cells, leucocytes, bacteria, fluke eggs, etc., according to the processes going on in the liver.

The hemorrhages, lack of sufficient gall, consequent disorder in digestion, pathological changes, etc., rapidly lead to a general cachexia, weakness, and emaciation.

In animals infested with flukes it has been noticed that the blood is poor in haemoglobin and that the number of blood corpuscles is below the normal.

For a more detailed discussion, see Schaper (1890).

As already stated, the symptoms and pathology here given are based chiefly upon observations made on sheep, but what has been said of the disease in sheep may also be said of the disease in cattle, except that the latter, on account of their greater strength, can better withstand the attack, and the symptoms are accordingly not so marked.

Diagnosis.—Flukes are said to be found in the faecal matter during the fourth stage, but their eggs may be found much earlier. Accordingly, if fluke disease is suspected a positive diagnosis may be made by

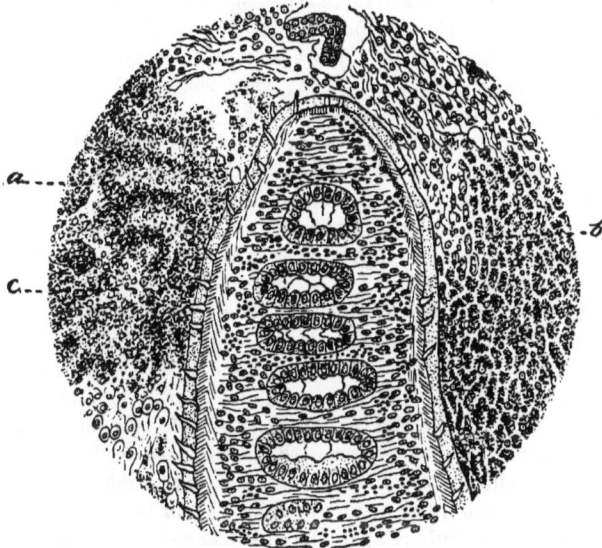

Fig. 16.—Drawing from a microscopic preparation showing a fluke in the tissue of the liver: *a*, necrotic liver tissue; *b*, atrophic liver cells; *c*, spines on the fluke, showing the outline of the body. (After Schaper, 1890, Pl. III, fig. 5.)

a microscopic examination of the faeces to find the ova. In order to do this it is often sufficient to place a minute portion of faecal matter on a slide, add a drop of water, and examine under a low-power lens.

An easy method of concentrating the eggs in a given amount of manure to be examined, so that the microscopic examination will be facilitated, is to place the faecal matter in a jar of water, shake well, filter through a wire net, and allow it to settle. The fluke eggs will settle on the bottom with the heavier matter, but a great deal of vegetable material will be caught by the wire netting or will float. The part which floats can then be drained off with water, leaving the eggs in the more solid matter, which can then be examined microscopically.

If facilities for a microscopic examination are not at hand, it is best to sacrifice one of the animals of the herd—the one in which the symptoms are most pronounced—and examine its liver for flukes.

Position of the parasites.—For the most part the flukes are confined to the gall ducts; some, however, are found in the parenchyma of the liver; a few reach the portal veins and cause endophlebitis, thrombosis, and emboli; others enter the liver veins and are carried to various parts of the body; upon passing the heart they reach the lungs, where they can give rise to hemorrhagic centers, canals with bloody contents, or even nodules. From the pulmonary arteries they could reach the pulmonary veins, and from there may be carried by the blood to any part of the body. The presence of flukes in peripheral portions of the body is, however, exceptional.

Influence of age.—It has been noticed in epizootics that calves and cattle under three years are more seriously affected by the disease than are older animals. This is undoubtedly due to the fact that the older animals are stronger, and hence are able to resist more.

It has, however, been shown that very young calves are comparatively rarely infested with flukes (see fig. 17); a fact which is easily understood when we recall that they are, from their mode of life, food, etc., less exposed to the infection than the older animals, which live almost entirely upon pasture, and, taking in a great amount of grass, naturally stand in danger of swallowing a greater number of the cercariae. Bulls which are kept close are generally free from these worms.

Geographical distribution; fluky years and fluky seasons.—This parasite has a very wide distribution, being found in Europe, Asia, Africa, North America, and South America. As a general rule, it can be said that the parasite is found on the lowlands—marshes, valleys, etc.—but is generally absent from the highlands; and this is in accordance with the facts observed in connection with the life history, for the intermediate host is a snail which lives in marshes and marshy districts, but is generally absent from the dry highlands. With this same general law of distribution, dependent upon the physical geography of the country, we can correlate two other general statements in regard to the occurrence of the parasite, and hence of the disease, based upon the humidity of the season—namely, fluke disease is more frequent in wet years ("fluky years") than in dry years, and fluke disease is more prevalent after the wet months of the year than after the dry months.

In wet years, namely, in years of heavy rainfall, the overflow of water naturally extends the limits of marshes and carries the snails over a greater area. Furthermore, the ground being more moist, the eggs have greater chances for development, and the infection is thus spread.

An idea of the frequency of the parasites during different months of the year may be obtained from an examination of fig. 17. On this chart Leuckart has plotted the animals slaughtered at the Berlin abattoirs during the years 1883–84, according to statistics furnished by Hertwig. The table covers 94,387 head of cattle and 77,848 calves. Of the cattle, about three-fourths to four-fifths were infested with flukes, and of these 3,428 were so badly infested that their livers were condemned. Of the calves, only 154 livers were condemned.

The table shows us that the parasites are present the entire year; also that there are two periods during the year, namely, from October to January (highest in October) and from March to April, in which the

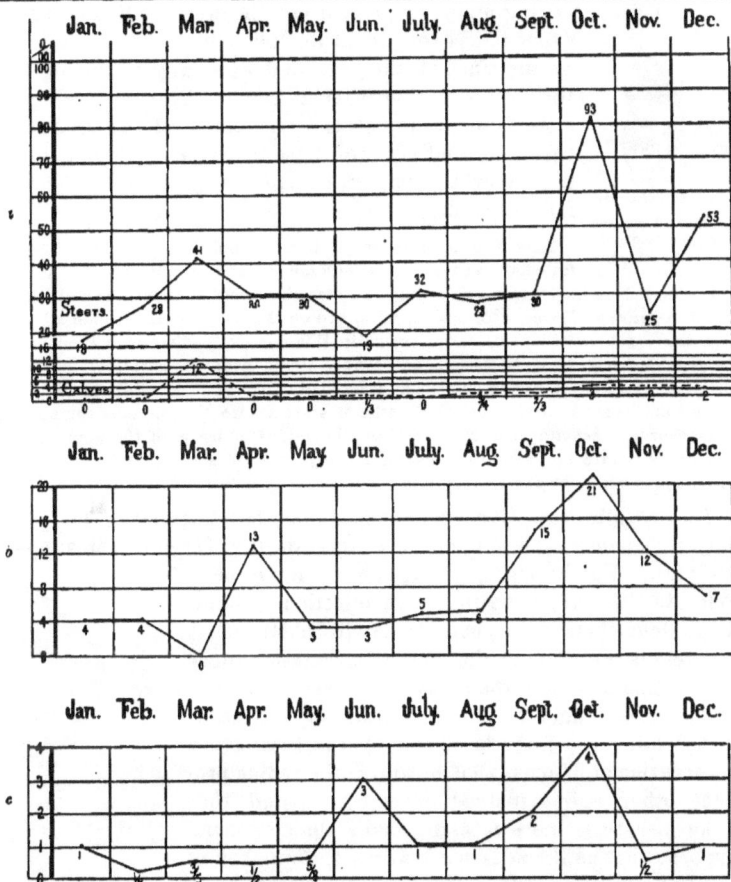

FIG. 17.—Tabular diagram of the occurrence of the Common Liver Fluke (*Fasciola hepatica*) during different months of the year: *a*, cattle; *b*, sheep; *c*, swine. (After Leuckart, 1889, p. 301, fig. 147.)

livers are particularly infested, or so altered as not to be fit for food. Leuckart has interpreted these figures as signifying that the winter maximum pointed to an infection in the fall, while the summer maximum pointed to one in early spring, namely, during the wettest seasons of the year.

According to Lutz (1892), Oahu and Kauai of the Sandwich Islands suffer considerably from fascioliasis. In some parts of Oahu nearly all the cattle have been destroyed by the disease; the sheep from dry districts, however, are not affected. Of 602 calves examined at Honolulu, 298 were found infested; of 2,186 cattle, 1,313 were infested, so that about four-sevenths of the animals were diseased.

In this country we have no exact statistics covering this parasite, but Francis, in writing upon the presence of the worm in Texas, states

"that it is exceptional to find a liver free from them at any time of the year, and especially so during the spring;" also, that "heifers coming 2 years old suffered more than at any other age. Many of the cattle and sheep die, and many of those that recover do not thrive the following summer, but remain poor and weak and fail to breed."

FIG. 18.—*Limnaea truncatula*, natural size and enlarged. (After Leuckart.)

Most American authors (Hassall and Francis excepted) have failed to recognize the difference between this species and *F. magna*, so it is in many cases impossible to determine whether an author had before him *F. hepatica* or *F. magna*, or both species, and on this account it is impossible to give the exact distribution of the worms in this country. That it is common in Texas is shown by Francis' article, and I have found the same parasite quite common (August, 1893) in Texan cattle slaughtered at Chicago. I have also found it in other than Texas cattle, although I can not state where the animals came from. Law records *F. hepatica* from sheep on Long Island. Curtice is of the opinion that *F. hepatica* is rare in the United States, but says that in sheep it is "reported by sheep books and newspaper articles."

Time of infection.—Gerlach supposed that the infection takes place only in summer and fall, but the diagram (fig. 17) does not support his view. Furthermore, young flukes have been found in February, pointing to an infection in January. Nevertheless, the general rule will hold true that in a temperate climate the time of greatest danger of infection is during the summer and early fall.

That the danger of infection gradually decreases in the fall and winter is shown by the interesting observation of Thomas that in winter the rediae produce other rediae instead of cercariae, and that fluke disease is more fatal to snails than to mammals, so that as the season advances the number of

FIG. 19.—*Limnaea peregra*, natural size and enlarged. (After Leuckart.)

cercariae in the fields must reach its maximum and then gradually decrease.

In a warm and moist climate the conditions favorable to infection will naturally persist longer than in a cold and dry climate.

Source of infection.—The snails[1] which form the intermediate host of this parasite must, because of the transmission of fluke disease, be included among the worst enemies of the stock raiser.

---

[1] For a more detailed account, see Stiles, 1894–95, pp. 303–313.

Leuckart and Thomas experimentally demonstrated the truth of Weinland's view, that in Europe the intermediate host for this fluke is a small swamp snail (*Limnaea truncatula*). Leuckart also showed that the rediae (but not the cercariae) would develop in the young of another species of snail (*L. peregra*), and quite recently Lutz (1892 and 1893) has shown that in Oahu and Kauai (Sandwich Islands) two other snails may serve in this capacity (*L. oahuensis* Souleyet and *L. rubella* Lea). In the case of *L. oahuensis*, Lutz states that "the infection can take place only in young specimens." None of these four very closely allied species are recorded for America, and yet we find *F. hepatica* in both North America and South America, so that we must either have on this continent some other species of snail which may act as intermediate host, or some of the species described in America must be identical with some of the above-named forms.

FIG. 20.—*Limnaea humilis*, natural size and enlarged. (After Binney.)

The forms which would especially fall under suspicion are *L. humilis* Say, in North America, and *L. viator* Orb., in South America.

FIG. 21.—*Limnaea oahuensis*, natural size and enlarged. (After Souleyet.)

This report is not the place to discuss the question as to whether these forms (*L. truncatula, L. peregra, L. oahuensis, L. rubella, L. humilis*, and *L. viator*) represent six well-established species or not, as that is a matter for conchologists to decide. Suffice it to say that specialists in conchology have described snails under these names; that the forms are all so very closely related that a zoologist would not commit a very grave offense against systematic zoology if he were to consider them as varieties of two or three species; that the forms described under the names *L. truncatula, L. oahuensis*, and *L. rubella* are known to serve as intermediate hosts for the parasite now under discussion; that in Europe the rediae (but not the cercariae) develop in *L. peregra*, and that it is probable, though not demonstrated as yet, that *L. humilis* is intermediate host for North America and *L. viator* for South America.

Treatment.—Hygiene must play a much more important rôle in the treatment of this disease than therapeutics, for while the knowl-

FIG. 22.—*Limnaea viator*, natural size and enlarged. (After d'Orbigny.)

edge of the life history of the parasite shows us how we may to a certain extent prevent the disease, no drug is known which can be relied upon to kill the flukes or dislodge them from their habitat. A great many drugs have been tried in the hope of accomplishing this end, but although some authors recommend the use of anthelminthics, most writers admit that such drugs are practically useless in this disease, and that the only treatment practicable is to use stimulants and tonics (various iron salts, walnut leaves, pepper in alcoholic drinks, calamus, etc.), with good nourishing food, such as lupine seeds, lupine hay, roasted malt, linseed cakes, oats, bran, etc., rich in protein, in order to build up the system and carry the animal through to the fourth stage of the disease, when the flukes will die or, as some authorities state, wander out spontaneously; and, in case the pathological lesions are not too great, the live stock will have an opportunity to recover. Many authors recommend astringents and diuretics (salt, juniper berries, turpentine, etc.) to meet the hydropic complications.

The following are some of the formulae given by various authors for fascioliasis in sheep, and the same medicaments may be used for this disease in cattle:

(1) The following is advised by Delafond. Make into a paste with water and allow to ferment, then bake in an oven. Give morning and evening. In about fifteen days this bread is said to produce improvement.

Mixture.	Metric.	Approximate equivalents.		
		Avoirdupois.	Apothecaries'.	Imperial troy.
Undressed wheat meal.	1 kilogram..	2⅕ pounds..........	2.7 pounds...... ...	2.7 pounds.
Oatmeal	2 kilograms.	4⅖ pounds..........	5.3 pounds	5.3 pounds.
Barley meal	1 kilogram..	2⅕ pounds..........	2.7 pounds	2.7 pounds.
Sulphate of iron.....	30 grams....	1 ounce 25 grains ...	463 grains..........	463 grains.
Carbonate of soda...	30 grams....	1 ounce 25 grains ...	463 grains..........	463 grains.
Table salt	200 grams...	7 ounces 24 grains ..	3,086 grains = 6⅜ ounces.	3,086 grains = 6⅜ ounces.

(2) The following is Hauber's lick for 100 sheep:

Mixture.	Metric.	Approximate equivalents.		
		Avoirdupois.	Apothecaries'.	Imperial troy.
Sulphate of iron.....	60 grams....	2 ounces 50 grains ..	926 grains = 1.7 ounces.	926 grains = 1.7 ounces.
Calamus root	500 grams...	17 ounces 279 grains.	7,716 grains = 1 pound 4 ounces.	7,716 grains = 1 pound 4 ounces.
Crushed oats	20 liters	21⅛ quarts, U. S	21⅛ quarts, U. S	17⅗ quarts, imperial.
Roasted barley malt .	20 liters	21⅛ quarts, U. S	21⅛ quarts, U. S	17⅗ quarts, imperial.

(3) The following is Hauber's lick for 50 sheep:

Mixture.	Metric.	Approximate equivalents.		
		Avoirdupois.	Apothecaries'.	Imperial troy.
Sulphate of iron...	30 grams.	1 ounce 25 grains ..	463 grains............	463 grains.
Powdered juniper berries.	500 grams.	17 ounces 279 grains.	7,716 grains = 1 pound 4 ounces.	7,716 grains = 1 pound 4 ounces.
Gentian...........	500 grams.	17 ounces 279 grains.	7,716 grains = 1 pound 4 ounces.	7,716 grains = 1 pound 4 ounces.
Grits	20 liters...	21⅛ quarts, U. S.....	21⅛ quarts, U. S.......	17⅗ quarts, imperial.

(4) The following lick for 300 sheep is highly indorsed by some authors, but not considered of much value by Zürn. A portion of this

mixture is given every other day for awhile, and then once every fourteen days through the summer.

Mixture.	Metric.	Approximate equivalents.	
		U. S. apothecaries', or wine measure.	Imperial troy.
Powdered lime	5 liters	5¼ quarts	4⅜ quarts.
Powdered table salt	10 liters	10½ quarts	8⅘ quarts.

(5) Mojkowski reports good results in treating sheep twice a day for a week with 0.7 to 1 gram (metric) of napthaline (= 7.7 to 15⅔ grains apothecaries' or imperial troy).

(6) Zürn suggests the following to be mixed and given to cattle in four doses in two days:

Mixture.	Metric.	Approximate equivalents.		
		Avoirdupois.	Apothecaries'.	Imperial troy.
Powdered wormwood	90 grams	3 ounces 76 grains	1,389 grains = 2.89 ounces.	1,389 grains = 2.89 ounces.
Powdered calamus root	90 grams	3 ounces 76 grains	1,389 grains = 2.89 ounces.	1,389 grains = 2.89 ounces.
Sulphate of iron	15 grams	½ ounce	231.5 grains	231.5 grains.

(7) Bunk advises 30 to 60 grams (=1 ounce 25 grains to 2 ounces 50 grains avoirdupois=463 to 926 grains apothecaries' or imperial troy) of benzine as a daily dose for each steer, to be given in mash.

The butcher's knife will be found a much more practicable means of treatment than any of the prescriptions given above, and the earlier in the disease that the animals are slaughtered the better condition they will be found in. In the early stages of the malady, as was seen above, there is a tendency on the part of the animals to fatten, due possibly to the increased flow of bile and the consequent acceleration in digestion, and, according to several authors, this fact has been taken advantage of by certain sheep dealers who have purposely exposed their flocks to fluke infection in order to fatten them early in the season.

In the case of cattle infected with *F. hepatica* it will scarcely be necessary to take such strenuous precautions as with sheep, for, as already stated, the disease is by no means as fatal to cattle as to sheep; in fact, in the vast majority of cases the presence of the parasites in cattle is not recognized until after the animals are slaughtered. This must not, however, be interpreted as meaning that the disease in cattle may be ignored, but merely that the disease in sheep must receive much more prompt attention than the disease in cattle.

If sheep are pastured in the same region as cattle, the presence of this parasite in cattle becomes doubly important, for in this case the disease will be spread to sheep and may cause heavy losses. Prompt measures to suppress the disease and isolation of the infested cattle should accordingly be resorted to.

Preventive measures.—For an excellent and more detailed account of the preventive measures, the reader is referred to Thomas (1883, pp. 296–305), of which the greater part of the following is a summary:

As seen from the life history of the parasite, four conditions are necessary for the propagation of this disease in any given district, namely: (1) The presence of fluke eggs; (2) wet ground, or water during the warmer weather, in which the eggs may hatch; (3) a snail (*L. truncatula,* or certain other species) which will serve as intermediate host; (4) herbivorous animals must be allowed to feed upon the infected pastures without proper precaution being taken to prevent infection. Destroy any one of these conditions and fluke disease will be destroyed; control any one of these conditions and the disease will be controlled in equal measure.

These conditions may be controlled or held in check by the following means:

(1) *To prevent the scattering of eggs in the fields:*

(*a*) In buying cattle or sheep, do not purchase any from a fluky herd, as they may introduce the disease to your farm.

(*b*) If animals are fluked, send those which are most affected to the butcher and place the others on dry ground.

(*c*) Destroy the livers of the slaughtered fluked animals, or if used as food for animals (dogs, etc.,) they should first be cooked in order to kill the eggs; if this precaution is not taken, the fresh eggs will pass through the intestine of the dogs uninjured and be scattered over fields.

(*d*) Manure of fluky animals should never be placed upon wet ground. It is, however, not dangerous to use such manure upon dry ground.

(*e*) "As rabbits and hares may introduce the disease into a district, or may keep up an infection if once introduced, these animals should be kept down as much as possible." This is not always practicable.

(*f*) Where animals very heavily infested with flukes have pastured on a given piece of ground, some one should go over the field with a spade and spread out the patches of manure, so that it will dry more rapidly, and thus the eggs may be more quickly destroyed. A spade full of lime or dust will aid in drying up the manure patches.

(*g*) Manure of fluky animals should not be stored where it can drain into pastures.

(2) *To control the second condition,* i. e., *marshy ground:*

(*a*) The marshes should be drained, if possible, so that the snails may be gotten rid of.

(*b*) It has been noticed that sheep which pasture on salty marshes are not fluked; accordingly dressings of salt, to which lime may be added, should be spread over the pasture, as salt and lime will destroy the embryos, the encysted cercariae, and the snails. May to August are the best months for scattering these substances.

Lime will destroy the grass for immediate use, but will in some cases be advantageous to the soil. The farmer must decide for himself whether he should use salt alone or lime and salt.

(c) If the marshy ground can not be controlled, place the animals on higher ground.

(3) *To destroy the snail.*—This may be done by draining the fields, thus depriving the snails of the conditions necessary for their development, or by the free use of salt and lime.

(4) *General precautions to be taken:*

(*a*) It is known that salt will kill the cercariae; accordingly if salt is given to the animals they stand a better chance of escaping hepatic infection, even if the germs are swallowed, not only because this substance kills the young flukes, but because it aids the animals in their digestion. The following experiment is interesting in this connection:

A number of uninfected sheep were selected and divided into two flocks, then placed upon pasturage which was known to be infected. One flock received no special attention, while the sheep of the other flock were fed a quarter of an ounce of common salt mixed with half a pint of oats every day that they were on the pastures; but when fed upon turnips, vetches, etc., the allowance of salt and corn [= oats] was not given. The first flock were so infected with flukes that they could not be kept through the winter, while the second flock was quite sound. The corn [= oats] and salt had cost about 3*s*. (75 cents) per head; the profit was about 50*s*. ($12.50) per head.—T. P. HEATH, Western Morning News, October 14, 1882.

(*b*) A daily allowance of dry food should be given.

(*c*) If fields are overstocked the animals will be obliged to graze very close to the ground, and will thus be more liable to become infected; accordingly, in order to prevent this close grazing, fields should not be overstocked.

(*d*) Animals should not be left too long upon the same pasture.

(*e*) Raised watertanks should be placed in the pastures so that the herds will not be forced to drink from pools, etc. As it is difficult for snails to get into such drinking tanks, there will be little fear of infections from tanks of this sort.

ABATTOIR INSPECTION.

Fluked animals as food.—If only a few flukes are found in the liver and these have not caused any extensive pathological changes, there seems to be no valid reason for condemning the entire organ as food, for the eggs would be perfectly harmless if eaten; the adult parasites, if swallowed alive, might cause some temporary injury, but as liver is well cooked in this country, there is scarcely any chance that the adult worm would be swallowed alive; if the pathological change is confined to a portion of the liver, that portion can be cut out and the rest may be used for food; in case of a general cirrhosis, or in case of suppurating inflammation of the tissue, caused by the wandering of flukes through the same, the liver should be condemned to the tank. There is generally no particular alteration to be noticed in the flesh of fluked cattle, unless the livers are very far gone, in which case the meat is more "flabby" and lighter than usual. In the case of badly fluked sheep, the flesh is of a very poor quality and contains but little nourishment; it is pale and "flabby," and according to European autnors it should not be placed on the market in case the sheep have passed Gerlach's second stage of the disease.

JURISPRUDENCE.

In this country we have no general laws protecting a person in case he buys fluked animals. In Germany, Austria, and Switzerland certain laws protect the buyer, so that if fluke disease shows itself in a flock within a stated time after purchase the contract is void.

THE COMMON LIVER FLUKE IN MAN.

This parasite is rare in man, only about twenty cases being on record of its presence in the bile ducts. It is not at all impossible that the parasites described as *Hexathyridium venarum*, *Distomum oculi-humani* (*D. ophthalmobium*), and *Monostomum lentis* are young erratic liver flukes.

The fluke may produce serious trouble in man, which may result fatally.

VARIETIES OF THE COMMON LIVER FLUKE.

Several varieties of the Common Liver Fluke have been described by different authors, and although they have not yet been recorded in this country, they should be mentioned briefly in this report:

Fig. 23.—The Narrow Liver Fluke (*Fasciola hepatica angusta*), natural size (original, from one of the cotype specimens).

(a) THE NARROW LIVER FLUKE (*Fasciola hepatica angusta*) OF SENEGAL CATTLE AND MAN (?).

[Figs. 23 and 24.]

This variety has recently been described by Railliet (1895, p. 338) from specimens taken from cattle slaughtered at St. Louis, Senegal. Blanchard (1895, p. 733) thinks it identical with *Fasciola gigantica* (see p. 49) of the giraffe. He also considers it identical with a parasite expectorated by a French naval officer in Brazil and recorded by Gouvea (1895). See also the next variety.

(b) THE EGYPTIAN LIVER FLUKE (*Fasciola hepatica aegyptiaca*) OF BUFFALO AND CATTLE.

[Figs. 25 and 26.]

This parasite was originally described by Looss (1896, pp. 33–36, 192) as a variety of the common fluke, but he has recently written to us that he is now inclined to look upon it as a distinct species. He found the parasite in the liver of cattle (*Bos taurus*) and buffalo (*Bos bubalis*). Blanchard (1896, p. 733) evidently considers this form identical with both the narrow fluke (*F. hepatica angusta*) and the giant fluke (*F. gigantica*, p. 49).

(c) THE COMMON LIVER FLUKE (*Fasciola hepatica caviae*) OF GUINEA PIGS.

SYNONYMY.—*Distomum caviae* Sonsino, 1890; *Fasciola hepatica* var. *cariae* (Sonsino) Sonsino, 1896.

Sonsino described this parasite from the guinea pig as a distinct species, but he now believes it to be a variety of the Common Liver Fluke.

3. The Giant Liver Fluke (*Fasciola gigantica*) of Giraffes, Cattle (?), and Man (?).

[Fig. 27.]

SYNONYMY.—*Fasciola gigantica* Cobbold, 1856; *Distomum giganteum* Diesing, 1858; *Distoma hepaticum* ex p. of Gervais and van Beneden, 1858; *Fasciola gigantea* Cobbold, 1859; *Cladocoelium giganteum* (Diesing) Stossich, 1892 ex. p.

BIBLIOGRAPHY.—For bibliography and technical discussion, see Stiles, 1894-95, pp. 139-143.

HOST. — Giraffe, cattle (?), and man (?). (See pp. 137-143.

This parasite was described from specimens taken in England from a giraffe belonging to a traveling menagerie. Blanchard (1895, p. 733) believes it identical with the narrow fluke (p. 48) and evidently also with the Egyptian fluke (p. 48).

FIG. 25.—The Egyptian Liver Fluke (*Fasciola hepatica aegyptiaca*), drawn from one of Looss' specimens, natural size (original). See p. 48.

4. The Large American Fluke (*Fasciola magna*) of Cattle and Deer.

[Figs. 28-35].

For anatomical characters, compare figs. 29 and 30 with key, p. 21.

VERNACULAR NAMES.—English, *The Large American Fluke, The Grand Fluke;* German, *Der grosse amerikanische Leberegel;* French, *Grand Distome;* Italian, *Distoma grande, Distoma magno.*

SYNONYMY.—*Distomum magnum* Bassi, 1875; *Fasciola carnosa* Hassall, 1891; *F. americana* Hassall, 1891;

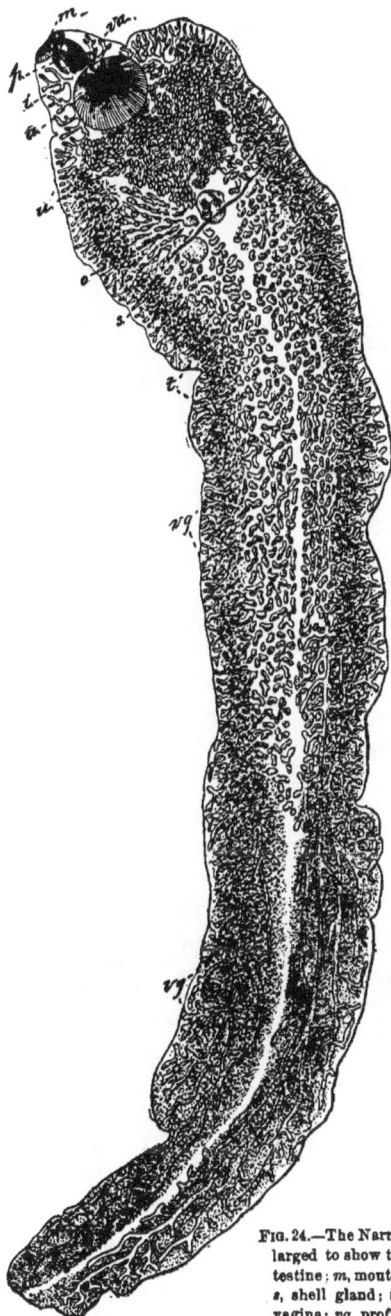

FIG. 24.—The Narrow Liver Fluke (*Fasciola hepatica angusta*), enlarged to show the anatomical characters: *a*, acetabulum; *i*, intestine; *m*, mouth with oral sucker; *o*, ovary; *p*, pharyngeal bulb; *s*, shell gland; *t*, profusely branched testicles; *u*, uterus; *va*, vagina; *vg*, profusely branched vitellogene glands. See p. 48.

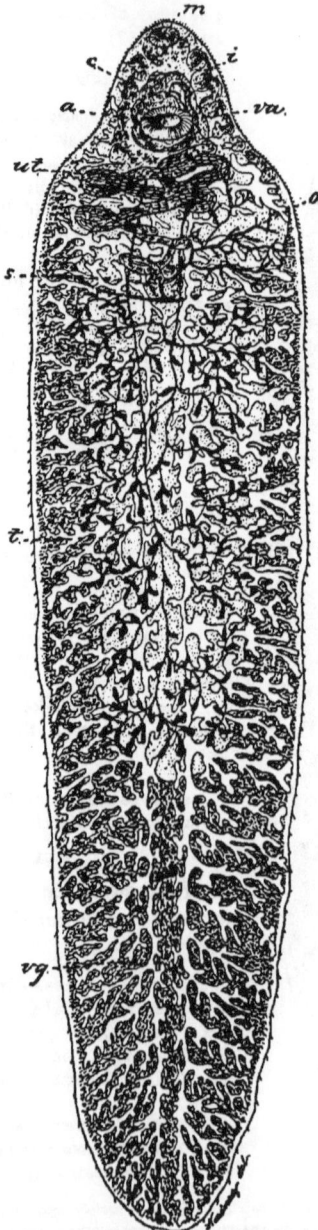

FIG. 26.—The Egyptian Liver Fluke (*Fasciola hepatica
aegyptiaca*), enlarged to show the anatomical characters: *a*, acetabulum; *c*, cirrus pouch; *i*, intestine;
m, mouth with oral sucker; *o*, ovary; *s*, shell gland;
t, profusely branched testicles; *ut*, uterus; *va*, vagina; *vg*, profusely branched vitellogene glands.
(After Looss, 1896, Pl. III, fig. 16.) See p. 48.

FIG. 27.—The Giant Liver
Fluke (*Fasciola gigantica*),
enlarged to show the anatomy. (After Cobbold,
1864.) See p. 49.

Distomum texanicum Francis, 1891; *D. americanum* (Hassall) Stiles, 1892; *Fasciola magna* (Bassi) Stiles, 1894.

BIBLIOGRAPHY.—For bibliography and technical discussion, see Stiles (1894-1895).
HOSTS.—Cattle, deer, and other animals. (See pp. 137-143.)
GEOGRAPHICAL DISTRIBUTION.—North America (Texas, Arkansas, Indian Territory, California, Iowa, Illinois, New York, and probably elsewhere); Europe (Italy).

The Large American Fluke appears to be more frequent in this country than the so-called Common Liver Fluke, although this opinion is the result of general impression from abattoir inspection rather than a view based upon actual statistics. The parasite was first described by Bassi, who found it producing a fatal epizootic among the deer of the Royal Park near Turin, Italy, where it is supposed to have been introduced with imported Wapiti from North America. Dinwiddie has found that in some counties of Arkansas practically all the cattle are infected with this worm, and for years the livers of cattle from certain districts have been unfit for use. As the infected area fell within the cattle-fever district, some persons

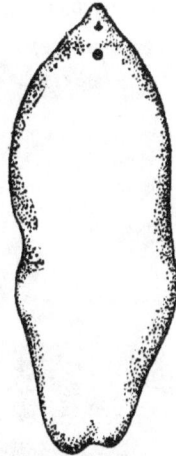

FIG. 28.—The Large American Fluke (*Fasciola magna*), natural size (original).

erroneously thought that the changes produced in the liver were due to Texas fever. Fortunately this species (so far as known) does not occur in sheep, and on that account it must be looked upon as of less importance than the common fluke.

Leidy (1891) thought this species identical with "*Distomum crassum*," which occurs in man, and Stossich (1892) considered it identical with the giant fluke *Cladocoelium giganteum* (=*Fasciola gigantica*) of giraffes.

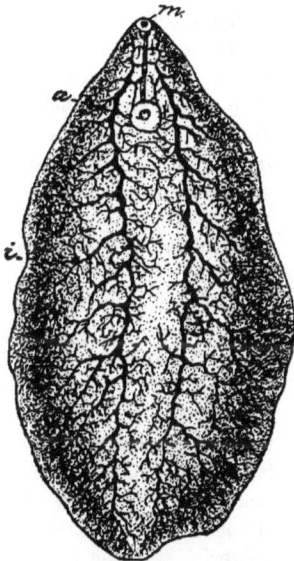

FIG. 29.—Macerated specimen of Large American Fluke, showing the digestive system and acetabulum. × 2. (After Stiles, 1894, p. 226, fig. 2.)

Life history.—The complete life history of this parasite has not yet been experimentally demonstrated, but as the species is so closely allied to the Common Liver Fluke, it will unquestionably be found that the life cycle agrees with that given for *Fasciola hepatica* (p. 30).

Upon several different occasions experiments have been instituted in this Bureau to trace out the life cycle, but the snails we have collected in the locality of the District of Columbia have thus far not taken the infection.

Egg.—The eggs (fig. 32) of *F. magna* can hardly be distinguished from those of *F. hepatica.* In general, however, they are slightly larger. In *F. hepatica* they vary from 0.105 mm. to 0.145 mm. (rarely 0.172 mm.) long by 0.063 mm. to 0.09 mm. broad. In *F. magna* they vary from 0.109 mm. to 0.168 mm. long by 0.075 mm. to 0.096 mm. broad. The structure of the egg agrees perfectly with that given for the Common Liver Fluke, so that a differential diagnosis in faecal examinations is impossible. Upon several different occasions we have raised the —

Miracidium (figs. 33, 34), which agrees with the ciliated embryo of *F. hepatica* (see p. 32). It is covered with a ciliated epithelium, and upon its anterior end is found a papilla in which an opening is perfectly visible. This opening leads into a thin string of tissue, evidently a rudimentary oesophagus, ending in a double-lobed body, which from homology with *F. hepatica* represents the rudimentary intestine. Immediately anterior of this is situated the ganglionic mass with the two cup-shaped eye-spots. In the posterior portion of the body a number of germ

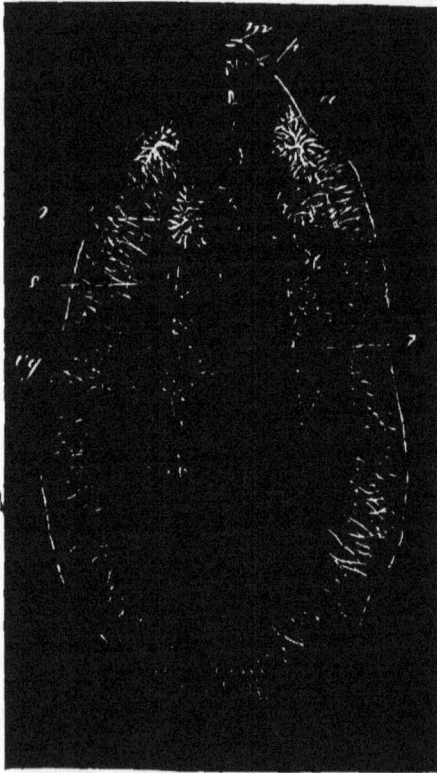

Fig. 30.—Macerated specimen of Large American Fluke (*Fasciola magna*), showing the anatomical characters: *a*, acetabulum; *m*, mouth with oral sucker; *o*, ovary; *p*, pharyngeal bulb; *s*, shell gland; *t*, profusely branched testicles; *vg*, profusely branched vitellogene glands. × 2. (After Stiles, 1894, p. 236, fig. 3.)

cells can be distinguished. The movements of this embryo agree with those of *F. hepatica.* The size varies according to contraction, but in general it may be given as 0.15 mm. long by 0.04 mm. broad.

Sporocyst, redia, and cercaria.—For a description of these stages (not yet known for *F. magna*), see pp. 32 and 33.

The disease.—The remarks upon this subject on page 36, under *F. hepatica*, will apply in a general way to this parasite also. The large fluke appears to be more dangerous for cattle, however, than the common fluke. According to Francis, many cattle die from the effects of the common fluke, and those which recover do not thrive the following summer, but remain poor and weak and fail to breed—remarks which if well founded for *F. hepatica* would apply in a still greater degree for *F. magna*. Heifers coming 2 years old suffered more than at any other age. It is stated that the large fluke has caused disease among the dairy cows in California, and Francis is said to have investigated an outbreak in Texas where the loss ran into hundreds of cattle. In the Italian outbreak, the disease corresponded with fluke disease of sheep, and reached its highest stage during the winter and spring. Bitting (1895) records fluke disease in cattle for Florida, but attributes it to the common fluke.

Symptoms.—See remarks under *F. hepatica*, pp. 34-36.

Pathology.—The pathological changes brought about by this form have never been studied in detail, but the earlier changes will doubtless agree with those described for *F. hepatica*. In heavily infected livers there is a much greater tendency to the formation of large cysts in the liver, in which several

FIG. 31.—A section of the cuticle of Large American Fluke (*Fasciola magna*), showing the spines. (After Stiles, 1894, p. 227, fig. 7.)

parasites are present. Dinwiddie has described a post-mortem as follows:

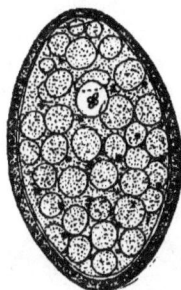

FIG. 32.—Egg of Large American Fluke, showing the germ cell, surrounded by a large number of vitelline cells, and an eggshell provided with a cap. (After Stiles, 1894, p. 227, fig. 4.)

Apparently in good health and fair butchering condition. The "fat caul" seen on first opening the abdomen as a large sheet was dotted with black spots and streaks. Lymphatic glands on the concave surface of the liver were much swollen and black in color. The liver itself was enlarged and darkened on the surface, with a number of prominent elevations, some appearing like blisters and some more or less solid, and varying greatly in size. A longitudinal section showed the presence of many cavities, some containing a dark fluid in which were floating granules and shreds of tissue. One very large cavity, about 2 inches in diameter, with irregular yellowish colored walls, besides the dark-colored fluid above mentioned, contained also two flat, leaf-like bodies about one inch in length and slightly less in breadth. They were fished out and recognized as "flukes." More of these were obtained from other cavities. Several other cavities contained solid, greenish-yellow, gritty matter, and no parasites. A section made through the liver in any direction cut through one or more of these cysts. They were situated near the surface of the organ or in its substance indiscriminately. Those that contained the "fluke" were usually of medium or smaller size, and the parasite was found folded or curled upon itself longitudinally and surrounded by fluid. * * * The shreds of tissue found in

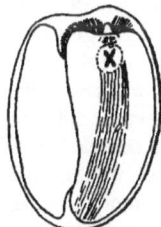

FIG. 33.—Ciliated embryo (miracidium) of Large American Fluke within the eggshell. (After Stiles, 1894, p. 227, fig. 5.)

those cysts, which did not contain the living parasites, were shown by microscopic examination to be the débris of dead and partly decomposed flukes.

Such were the gross appearances of the livers of at least three-fourths of the cattle slaughtered during the spring and summer at this place, and about 90 per cent of all coming from certain ranges in St. Francis and Lee counties [Arkansas].

FIG. 34.—Free embryo (miracidium) of Large American Fluke (*Fasciola magna*), showing ciliated epithelium, boring papilla, rudimentary oesophagus, and intestine; eye-spots situated above the ganglionic mass, and germ cells. (After Stiles, 1894, p. 227, fig. 6.)

I am inclined to think that the effects of the parasite upon cattle have possibly been somewhat over-estimated, for I have seen cattle in abattoirs which were apparently in excellent condition and yet whose livers were literally composed almost entirely of flukes and their cysts. The question arises whether other factors (Texas fever, blackleg, etc.) also were not concerned in the outbreaks among cattle which have been attributed to this parasite. A small number of these worms certainly has little or no appreciable effect upon cattle, and even when a large number is present the effects do not appear to be very great in the case of full-grown steers. The fact that Francis has attributed the death of a number of cattle (chiefly 2-year-olds) to this worm is deserving of attention, but the published accounts of these outbreaks are not detailed enough to allow a satisfactory conclusion. While it must be admitted that the pathological changes in the liver of cattle caused by this parasite can not help producing some effect upon the host, we are perfectly warranted in the statement that the Large American Fluke bears a much less important relation to the cattle industry than the common fluke bears to the sheep industry.

Diagnosis.—Same as for the common fluke. (See p. 39.)

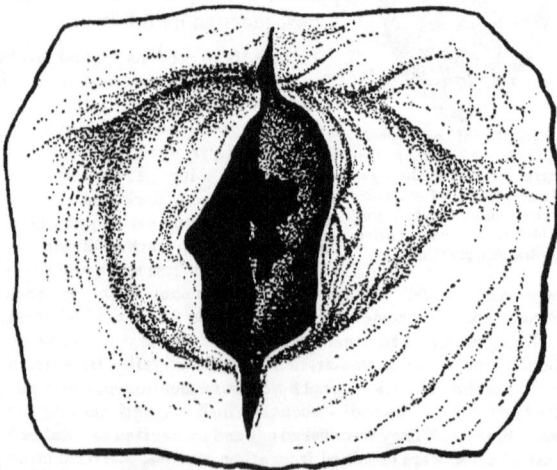

FIG. 35.—Cyst in the liver, caused by Large American Fluke. (After Stiles, 1894, p. 226, fig. 1.)

Position of the parasites.—Thus far the large fluke has been recorded only in the liver and lungs.

Influence of age.—The remarks on page 40 under this heading will doubtless be found to hold for the large fluke also.

Geographical distribution; fluky years and fluky seasons.—F. magna is known from the localities given on page 51. (See also remarks under this head in the discussion of *F. hepatica*, p. 40.)

Time of infection.—The remarks under this head on page 42 will apply in a general way to this parasite also.

Source of infection.—The intermediate host is as yet unknown, but it should not be a difficult matter to determine this point in the infected areas. It will undoubtedly be found to be a snail, probably of the genus *Limnaea*.

FIG. 36.—Lancet Fluke (*Dicrocoelium lanceatum*), natural size (original).

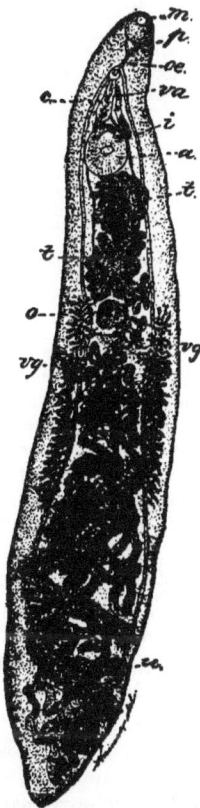

Treatment and preventive measures.—See pages 43–47.

ABATTOIR INSPECTION.

Fluked animals as food.—Regarding the flukes in the liver, see page 47. I have examined the meat of a large number of cattle whose livers were infested with this parasite, and have been unable to find any ground for excluding the meat from market. (See also pp. 47–48.)

DICROCOELES (Distomes of the Genus Dicrocoelium).

One representative of this genus, namely, *D. lanceatum*, has been recorded for cattle, sheep, and hogs, and a second species (*D. pancreaticum*) has been recorded for cattle and sheep, but there is no satisfactory evidence that either parasite is present in this country. (See p. 56.)

5. The Lancet Fluke (*Dicrocoelium lanceatum*) of Cattle, Sheep, and Swine.

[Figs. 36–39.]

FIG. 37.—Lancet Fluke, enlarged to show the anatomical characters: *a*, acetabulum; *c*, cirrus pouch; *i*, intestine; *m*, mouth with oral sucker; *o*, ovary; *oe*, oesophagus; *p*, pharyngeal bulb; *t*, lobate testicles; *u*, uterus; *va*, vagina; *vg*, vitellogene glands. (After Stiles & Hassall, 1894, Pl. IV, fig. 19.)

For anatomical characters, compare fig. 37 with key, p. 21.

VERNACULAR NAMES.—English, *Lancet Fluke*; German, *Der lanzettförmige Leberegel, das lanzettförmige Doppelloch;* French, *Distome lanceolé;* Italian, *Distoma lanceolato.*

SYNONYMY.—*Fasciola lanceolata* Rudolphi, 1803 [nec Schrank, 1790]; *Distoma lanceolatum* (Rudolphi) Mehlis, 1825; "*Distoma (Dicrocoelium) lanceolatum* Mehlis" of Dujardin, 1845; "*Distomum lanceolatum* Mehlis" of Diesing, 1850; "*Dicrocoelium lanceolatum* Dujardin" of Weinland, 1858; "*Fasciola Buchholzii* Jördens, 1801," misprint of Braun, 1889; *Dicrocoelium lanceatum* Stiles & Hassall, 1896.

BIBLIOGRAPHY.—No extensive bibliography as yet published. For detailed technical discussion, see Leuckart (1889, pp. 359–399).

HOSTS.—Man, cattle, sheep, swine, and other animals. (See pp. 137–143.)

GEOGRAPHICAL DISTRIBUTION.—Very extended, especially in Europe, but apparently not in England or North America.

Life history.—The complete development of this parasite is not yet known, although it is undoubtedly an indirect development with change of host, the intermediate host being some mollusk.

Von Willemoes-Suhm looked upon *Planorbis marginatus* as intermediate host; a *Limnaea* has also been viewed with suspicion.

FIG. 38.—Egg of Lancet Fluke (*Dicrocoelium lanceatum*) with contained embryo. × 700. (After Leuckart, 1889, p. 379, fig. 171.)

Leuckart found some cercariae in a *Planorbis*, which he suspected for a while represented the larval stage of this worm; but as none of these snails have as yet been experimentally proven to be the host of the larval stage of the parasite, the question is at present far from being solved. In Leuckart's most recent experiments he has fed eggs of the parasite to certain small slugs and noticed that the embryos escaped from the egg, but were unable to develop into the sporocyst stage. That the embryos escaped from the eggshell in the intestine of slugs points to the fact that the experiments are in the right direction, and that it will probably be found that some snail belonging to the family Limacidae—the slugs—in the order of the Pulmonata serves as intermediate host to the Lancet Fluke.

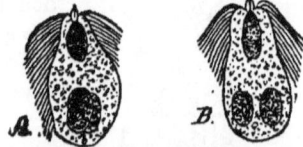

FIG. 39.—Free embryo (miracidium) of the Lancet Fluke: *A*, lateral view; *B*, dorsal view. (After Leuckart, 1889, p. 385, fig. 175 *A*, *B*.)

The Lancet Fluke is much less dangerous, owing to its smaller size and unarmed cuticle, than either the common fluke or the large fluke; and the pathological changes caused by the Lancet Fluke, even when present in large numbers, are scarcely ever more than a catarrhal affection of the gall ducts, rarely with secondary troubles. The parasite is frequently found in very large numbers, cases being recorded where 1,000 specimens or more have been taken from a single liver; it may occur alone or in company with *F. hepatica*. It has been recorded about six times in man.

Leidy (1856, p. 43) says that this parasite is "stated to be frequent in sheep in several of the Western States." This "statement" may be correct, but we have not yet been able to verify it.

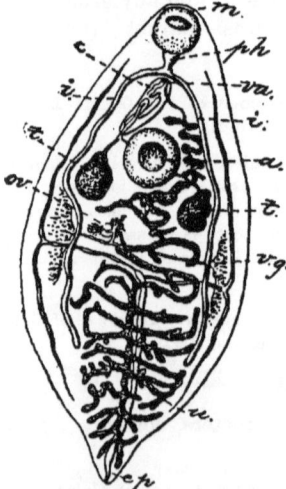

FIG. 40.—The Pancreatic Fluke (*Dicrocoelium pancreaticum*), enlarged to show the anatomical characters: *a*, acetabulum; *c*, cirrus pouch; *cp*, excretory pore; *i*, intestine; *m*, mouth with oral sucker; *ov*, ovary; *ph*, pharyngeal bulb; *t*, testicles; *u*, uterus; *va*, vagina; *vg*, vitellogene glands. (After Railliet, 1897.) See p. 57.

ABATTOIR INSPECTION.

In abattoir inspection, the rules given for infection with *F. hepatica* (p. 47) would apply to cases of infection with the Lancet Fluke.

6. The Pancreatic Fluke (*Dicrocoelium pancreaticum*) of Cattle and Sheep.

[Fig. 40.]

For anatomical characters, see key, p. 21.

SYNONYMY.—*Distoma pancreaticum* Railliet, 1890; *Distoma coelomaticum* Giard & Billet, 1892; *Distomum pancreaticum* Janson, 1893; *Distoma (Dicrocoelium) coelomaticum* Giard & Billet of Railliet, 1896; *Dicrocoelium pancreaticum* (Railliet) Railliet, 1897.

BIBLIOGRAPHY.—Railliet, 1897, pp. 371–377.

HOSTS.—Japanese cattle and sheep, Cambodia cattle and Indian buffalo. (See pp. 137–143.)

The Pancreatic Fluke, which is somewhat smaller than the common fluke but larger than the Lancet Fluke, has been found in Japan, Tonkin, and Cochin China, but is not yet recorded for North America; it is said to be present in about 50 per cent of the cattle and buffaloes of Cochin China, slaughtered in good condition, and in 90 per cent of the cachectic animals; it is found at all seasons of the year, both wet and dry. Its normal seat is the Ductus Wirsugianus and its branches, which are occasionally given a sausage-like appearance by the presence of the parasites. The local lesions developed by the presence of the Pancreatic Fluke are not generally very extensive; in many cases the pancreas seems quite normal; when the infection is extensive, however, this organ is thicker and heavier than usual; occasionally blackish streaks are noticed on the surface, representing the infected canals, but usually it is necessary to cut into the organ in order to recognize an infection. Even when the infected canals assume a sausage-like or moniliform appearance, no abnormal fluid appears to be present, and the thickening and induration of the walls are scarcely noticeable.

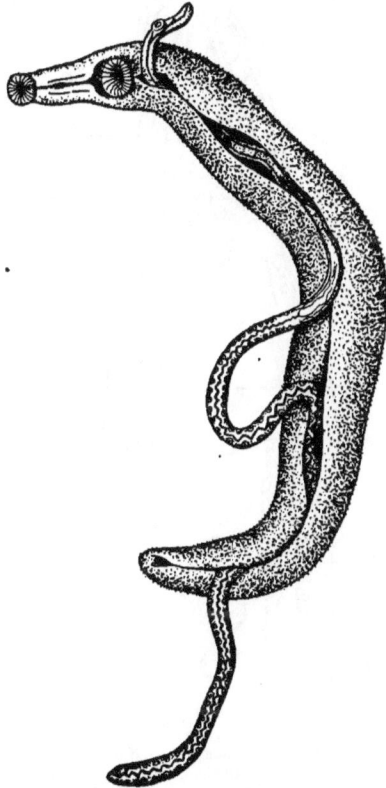

FIG. 41.—Male and female specimens of the Human Blood Fluke (*Schistosoma haematobium*), enlarged. × 12. (After Looss, 1896, Pl. XI, fig. 107.) See p. 58.

There is at present no reason to assume that these parasites would continue to live for any length of time if accidentally eaten by man; in fact, their direct transmission from cattle to man through eating sweet-breads infected by them is contrary to analogy.

Dioecious Distomes (Flukes of the Subfamily *Schistosominae*).

BLOOD FLUKES (Distomes of the Genus Schistosoma).

Flukes of this genus, only a few of which are known, live in the veins of mammals and birds. At least one species (*S. bovis*) is found in cattle and sheep, while the occurrence of a second form (*S. haematobium*) in cattle is as yet in need of confirmation.

The Human Blood Fluke has been found twice in this country: once in a foreigner on the "Midway" during the World's Fair, and once in New York.

7. The Human Blood Fluke (*Schistosoma haematobium*) of Man and Cattle (?).

[Figs. 41-44, 48.]

For anatomical characters, compare figs. 41-43 with key, p. 21.

SYNONYMY.—*Distomum haematobium* Bilharz, 1852; *Schistosoma haematobium* (Bilharz) Weinland, 1858; *Gynaecophorus haematobius* (Bilharz) Diesing, 1858; *Bilharzia haematobia* (Bilharz) Cobbold, 1859; (?) *Bilharzia magna* Cobbold, 1859; *Thecosoma haematobium* (Bilharz) Moquin-Tandon, 1860; *Distoma capense* Harley, 1864, nomen nudum; *Bilharzia capensis* Harley, 1864; *Bilharzia haematobia hominis* Kowalewski, 1895; (?) *Bilharzia haematobia magna* (Cobbold) Kowalewski, 1895; *Schistosomum haematobium* (Bilharz) of Blanchard, 1895.

BIBLIOGRAPHY.—For bibliography, see Huber (1894, pp. 294-305). For detailed anatomical study, see Looss (1895, pp. 1-108) and Leuckart (1894, pp. 464-534).

HOSTS.—Man, Sooty monkey (?), and cattle (?). (See pp. 137-143.)

GEOGRAPHICAL DISTRIBUTION.—Africa.

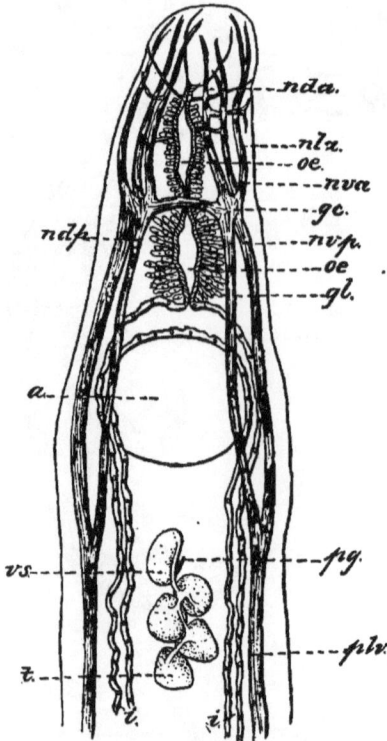

FIG. 42.—Anterior portion of male Human Blood Fluke (*Schistosoma haematobium*), showing the anatomical characters: *a*, acetabulum; *gc*, cerebral ganglion; *gl*, glands of oesophagus (*oe*); *i*, intestine; *nda*, dorsal anterior nerve; *ndp*, dorsal posterior nerve; *nla*, lateral anterior nerve; *nva*, ventral anterior nerve; *nvp*, ventral posterior nerve; *plv*, lateral posterior nerve; *pg*, genital pore; *t*, testicles; *vs*, vesicula seminalis. (After Looss, 1895, Pl. II, fig. 18.)

Life history.—The following may be taken as a summary of our present incomplete knowledge of the life history of this parasite. The eggs which are passed in the urine contain a ciliated embryo possessing a terminal papilla; a rudimentary intestinal sac, at each side of which is a large glandular cell; a rudimentary nervous system; an excretory system, and a number of germinal cells. While in the fresh urine the embryo is comparatively quiet, but more active movements can be brought about by the addition of water; water also causes the shell to burst, the embryo becoming free; preserved in urine the embryos die within about two hours. From this embryonic stage to the time when the parasites are found in the body we have no positive data concerning the life history, although clinical observation and analogy point to unfiltered water as the source of infection.

Sonsino concluded from his recent investigations that fresh-water crustaceans (*Gammarus Simoni*) and insects form the intermediate hosts; that the embryo develops through a larval stage ("*Dicotyle*"), but without an alternation of generations, and that man becomes infected by swallowing the "*Dicotyle*."

While analogy points directly to some fresh-water invertebrate as the intermediate host, the presence of germ cells in the miracidium points to a necessary alternation of generations as opposed to Sonsino's idea of a metamorphosis.

Hosts.—The Human Blood Fluke is found in man in Africa, especially in Egypt. A parasite found by Cobbold in the Sooty monkey (*Cercopithecus fuliginosus*), and described as *Bilharzia magna*, may possibly be identical with this parasite of man.

Cobbold, Leuckart, and Blanchard admit the identity of the two forms, while some other authors do not consider the point as yet established.

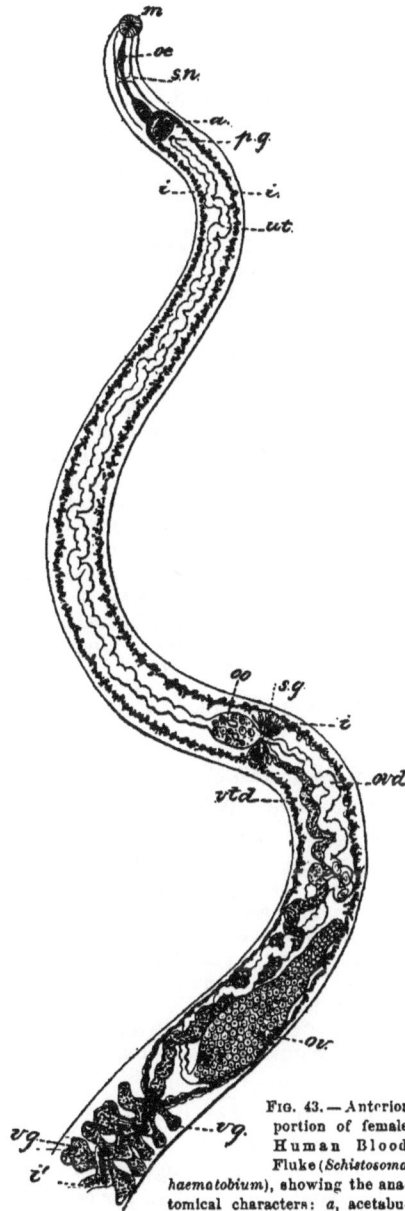

Fig. 43.—Anterior portion of female Human Blood Fluke (*Schistosoma haematobium*), showing the anatomical characters: *a*, acetabulum; *i*, intestine which is double for some distance, but the two caeca unite (*i*) back of the ovary; *oe*, oesophagus; *oo*, ootyp; *ov*, ovary; *ovd*, oviduct; *pg*, genital pore; *sg*, shell gland; *sn*, nervous system; *ut*, uterus; *vtd*, vitello duct; *vg*, vitellogene glands. × 38. (After Looss, 1896, Pl. XI, fig. 108.)

Schistosoma haematobium has been recorded once from cattle in Calcutta, but the determination is perhaps open to question. The exact data given are as follows:

Bomford found the peculiar uncinate ova of *Bilharzia* on microscopic examination of the large intestines of two Calcutta transport cattle destroyed on account of their being considered affected with rinderpest. In one case numerous eggs were found in a small portion of the caecum preserved in absolute alcohol. They were most numerous within or between the tubular glands of the mucous membrane, but were also present in considerable numbers in the submucous tissue below the muscularis mucosae. The alcohol had shriveled up the contents of the eggs, but the external form of the shell was preserved and the characteristic hook very clearly seen. In another bullock the ova were found in some papillomatous growths removed from the margin of the anus. In this case the form of the embryo in the ova could be distinguished. The ova exactly resemble those of *Distomum* (*Bilharzia*) *haematobium* hitherto found only in man (or a monkey), and in Africa, Arabia, or Mauritius. Sonsino's *Bilharzia boris* of Egyptian cattle differs in the spindle shape of its eggs and in their short, broad, caudate spine. These bullocks had not served in Egypt, but may possibly have obtained the parasites from Indian transport cattle which had done so. This parasite should be sought for in cases of Haematuria of cattle and when the ileocaecal ring is found congested. (See Mem. Med. Officers Army India, II (1886), 1887, p. 53.)

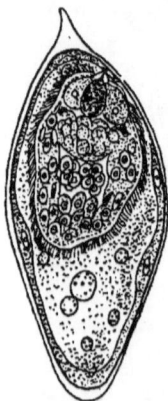

Fig. 44.—Egg of Human Blood Fluke (*Schistosoma haematobium*), with contained embryo, passed in the urine. × 285. (After Looss, 1896, Pl. XI, 112.)

8. The Bovine Blood Fluke (*Schistosoma boris*) of Cattle and Sheep.

[Figs. 45-47.]

For anatomical characters, compare figs. 45 and 46 with the key, p. 21.

SYNONYMY.—*Bilharzia bovis* Sonsino, 1876; *Bilharzia crassa* Sonsino, 1877; *Gynaecophorus crassus* (Sonsino) Stossich, 1892; *Gynaecophorus boris* (Sonsino) Railliet, 1893; *Bilharzia haematobia crassa* (Sonsino) Kowalewski, 1895; *Schistosomum boris* (Sonsino) R. Blanchard, 1895.

BIBLIOGRAPHY.—For bibliography, see R. Blanchard (1895, p. 191). For anatomical discussion, see Leuckart (1894, pp. 464-534).

HOSTS.—Cattle and sheep. (See pp. 137-143.)

GEOGRAPHICAL DISTRIBUTION.—Egypt, Italy, Sicily, India (?).

Fig. 45.—The Bovine Blood Fluke (*Schistosoma boris*), male and female. × 9. (After Leuckart, 1894, p. 467, fig. 204*A*.)

This parasite was discovered by Sonsino (1876) in Egypt in the portal veins of the ox and later he found it in sheep, while Grassi and Rovelli afterward found it in about 75 per cent of the sheep slaughtered at Catania, Sicily. The sheep were born and raised on the neighboring plains.

The worm is said to bring about in cattle and sheep the same lesions of the bladder, intestine, etc., which *S. haematobium* causes in man. Nothing is known regarding the life history.

THE DISEASE BILHARZIOSIS.

As this disease in man has been subjected to much more thorough study than the same malady in cattle and sheep, the human subject may well be taken as basis for the discussion.[1]

Source of infection.—As already stated, clinical observation and analogy point to unfiltered drinking water as the source of infection.

Position of the parasite.—The worms are found in the veins of the abdomen, the vena porta, vena linealis, vena renalis, and the venous plexus of the bladder and of the rectum.

Symptoms.—The *period of incubation* has not been definitely determined, but Hatch records the case of a patient who remained fourteen days at Suez and suffered from bilharzian haematuria one month after his arrival at Bombay. The young parasites appear to do no injury; in fact, even the adult worms seem to be inoffensive in themselves. The eggs on the other hand, armed with a sharp point, are the exciting cause of the disease. The position of the parasite in the venous system and the consequent location of the agglomeration of eggs determine the particular symptoms. Either the genito-urinary system is attacked, in which case haematuria is one of the first symptoms; or the large intestine is attacked and blood is noticed in the stools.

Fig. 46.—Cross section of Bovine Blood Fluke (*Schistosoma bovis*), showing the position of the female in the gynaecophoric canal. × 200. (After Leuckart, 1894, p. 472, fig. 209.)

If the parasites are lodged in the venous plexus of the genito-urinary system, the chief symptoms are: Haematuria; pains in the lumbar region, the left iliac fossa, the thigh, or in the vulva, which may be spontaneous or may accompany micturition; cystitis; vesical calculus; urinary fistulae; vaginal verminous tumors; nephritis.

The eggs accumulate in the capillaries, which they rupture; they traverse the mucosa and fall into the bladder, thus causing more or less hemorrhage; in this way the *haematuria* is established, which is often the initial symptom. At first the urine is quite bloody, but it gradually becomes clearer, and it is only at the end of micturition that muco-purulent flakes are expelled, in which numerous eggs and even embryos are found; the urine contains also epithelial cells, more or less pus, eggs, and occasionally embryos. On micturition sharp pains are felt at the base of the penis or at the gland, possibly due to the passage of eggs. The passage of eggs

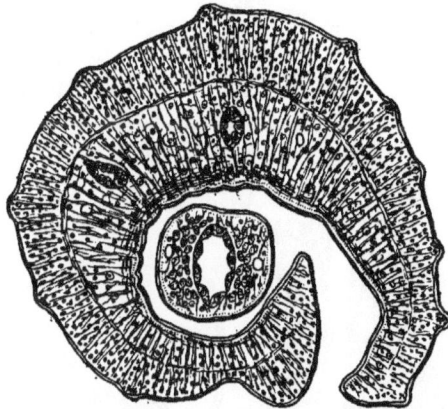

---

[1] This discussion is based chiefly upon Blanchard, 1895, pp. 69–93.

through the walls of the bladder give rise to *cystitis;* blood becomes more abundant in the urine after fatigue, coitus, or after taking alcoholics; clots may form and cause retention of urine; chronic urethritis may develop, evidently due to the presence of the eggs. In Egypt 80 per cent of the cases of *vesical calculus* coincide with bilharziosis; the formation of the calculi evidently results from the presence of the eggs, for the central nodule always contains one or more of these structures. Urinary fistulae, opening on the perineum, more rarely into the rectum, occasionally form. In women, the vagina may become the seat of a chronic inflammation; it is painful to the touch, exudes a bloody foetid discharge, and may become ulcerated or may be covered with numerous sessile or pedunculate *tumors*, which are very vascular and spongy, and contain the parasites or their eggs. The mucosa of the vagina, also the uterus and bladder, become impregnated with calcareous salts. *Nephritis* develops in grave cases.

If the parasites lodge in the veins of the rectum the lesions caused are analogous to those described for the genito-urinary tract; among the most prominent symptoms are bloody stools, dysenteric diarrhoea, enterorrhagia, and prolapse of the rectum. The mucosa is studded with numerous papilliform outgrowths, which occasionally attain considerable size and require surgical interference.

The heart, lungs, and liver generally remain normal.

Pathology.—The bladder is reduced in size, while its wall is greatly thickened, due chiefly to the hypertrophy of the muscularis; the mucosa is also thickened, and at certain points it is indurated by uric or calcareous deposits, but the principal lesion consists in ulcerations covered with sanious pus; the mucous membrane is infiltrated with numerous leucocytes, but with few eggs; the submucous connective tissue, however, contains numerous eggs, which also fill the blood vessels; most of the eggs

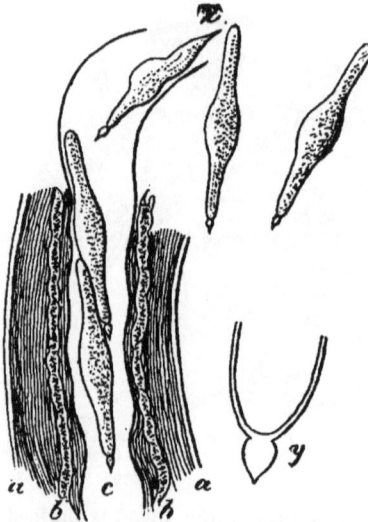

Fig. 47.—Eggs of Bovine Blood Fluke (*Schistosoma bovis*), showing the peculiar process on the end: *a, b*, layers of the oviduct; *c*, eggs in the oviduct × 180; *x*, eggs deformed by pressure; *y*, spinous process on end of egg × 700. (After Sonsino.)

are dead and more or less degenerated; frequently the mucosa is hypertrophied in places so that papillae are formed which are larger than those found in cases of simple catarrh of the bladder; they occasionally attain a finger in length, and may be recognized during life by cystoscopic examination. Harrison noticed in four cases out of five at Alexandria, Egypt, that the tumors developed in the tissue of the bladder had a carcinomatous character. Lesions analogous to those of the bladder are also observed in the lower third of the ureters and may extend as high as the kidney; the ureter is enlarged and tortuous; the mucosa irregular; its lumen may remain nearly normal in size, but its wall becomes very thick. The flow of the urine may be obstructed. The kidney increases in size, its calyx dilates, the division between cortical and medullary substance becomes indistinct, and the renal tissue may be reduced to an almost homogenous layer 3 to 4 mm. thick; miliary abscesses form on the surface; in short, a veritable hydronephrosis obtains, which results in atrophic lesions of the kidney and may finally end fatally; death may occur rather frequently from albuminuria; in the less grave cases the renal affection consists of a simple inflammation and parenchymatous nephritis; renal calculi may form; the organ may become the seat of a more or less intense cirrhosis. The vesicula seminalis and prostata may also contain eggs, and become more or less hypertrophied.

The polyps of the rectum, mentioned above, may attain 10 mm. to 13 mm. in length; the eggs are accumulated, especially in the mucosa, and may form masses 1.25 mm. thick, visible to the naked eye; a microscopic examination of the growths shows that they are composed in great part of mucosa, the glands (the normal length of which is about 0.5 mm.) becoming 2 to 3 or 3.5 mm. in length by 60 μ to 80 μ in diameter. Between the polyps the mucosa shows the lesions of chronic dysentery; all the tunics exhibit traces of a slow phlegmatic process; the submucosa is infiltrated with leucocytes; the muscularis may hypertrophy to three or more times its normal thickness.

The mesenteric lymphatic glands may hypertrophy, their substance becoming tumified, presenting small hemorrhagic centers, and containing eggs. The liver may contain eggs and become somewhat cirrhotic; the eggs accumulate in the branches of the portal veins, or after piercing the walls they lie in the hepatic parenchyma. The lungs may also contain eggs, as was shown by Mackie in the case of a patient who succumbed to pyemia following a purulent cystitis. He found in the lungs a large number of small metastatic abscesses limited by a necrotic tissue and containing a sanious pus with *Schistosoma* ova.

Diagnosis.—The diagnosis may easily be made by a microscopic examination of the urine to determine the presence of the egg.

Prognosis, etc.—The severity of the disease varies directly with the number of parasites (and hence the number of eggs) in the body. Fortunately, in the majority of cases the number of parasites is small, though it may increase from repeated infections to 500 or more. In cases of comparatively light infection, the disease is reduced to a slight chronic cystitis, with now and then exacerbations, in course of which a slight amount of blood and pus is passed in the urine. The disease may last for years without apparent increase. In the most severe cases death may occur from various causes; a rupture of the bladder, ascending pyelonephritis, uremia, albuminuria; the patient may die in marasmus, being exhausted by the dysentery or the anaemia.

Fig. 48.—Ureter of an Egyptian, with numerous uric-acid concretions, as a result of blood-fluke infection. (After Leuckart, 1894, p. 528, fig. 231.)

Bilharziosis is accordingly not such a fatal disease as has sometimes been supposed.

Prevention.—Avoid unfiltered or unboiled water in contaminated districts.

Treatment.—No experiments in treating cattle for this disease have been recorded.

In human practice Fouqnet seems to have had good success with capsules of extract of male fern; he begins with one capsule per day, afterward increasing the dose to two or in some cases to three capsules. The dosing is continued with persistency until the patient seems recovered; the dose is then reduced to one capsule daily

for one month. Intravesical injections of bichloride of mercury 1:5000 are advised in severe cases. Nitrate of silver, carbolic or boric acid are also used as injections or enemata. Napier claims good results with salicylate of soda; 40 grains before retiring. Surgical intervention is occasionally necessary in cases of severe lesions.

ABATTOIR INSPECTION.

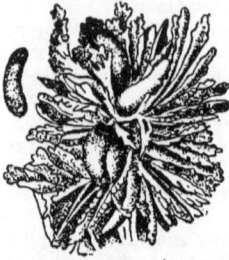

Fɪɢ. 49.—Conical amphistomes (*Amphistoma cervi*) in the rumen; tubercles from which the parasites have loosened. (After Railliet, 1893, p. 376, fig. 249.)

At present the blood flukes do not play any rôle in the inspection at American abattoirs. Should the parasite appear in this country it will probably first be found in Southern cattle, and the affected organs should be condemned in order to prevent the spread of the worm. There would, however, be no danger of transmission of the parasite from cattle direct to man.

AMPHISTOMES (Flukes of the Family Amphistomidae).

Of this family of worms, characterized by the position of the acetabulum in the posterior portion of the body, only one species (*Amphistoma cervi*) has as yet been recorded in the herbivorous animals of North America.

TRUE AMPHISTOMES (Flukes of the Genus Amphistoma).

9. The Conical Fluke (*Amphistoma cervi*) of Cattle and Sheep.

[Figs. 49-55.]

For anatomical characters, compare figs. 49 and 50 with key, p. 21.

Synonymy.—*Festucaria cervi* Zeder, 1789; *Fasciola cervi* (Zeder) Schrank, 1790; *Fasciola elaphi* Gmelin, 1790; *Monostoma elaphi* (Gmelin) Zeder, 1800; *Monostoma conicum* Zeder, 1803; *Amphistoma conicum* (Zeder) Rudolphi, 1809; *Amphistomum conicum* (Zeder) of Diesing, 1850; *Strigea cervi* (Zeder) Railliet, 1893.

Bibliography.—For bibliography, see Otto (1896, pp. 97, 98). For technical discussion, see Otto (1896), Looss (1896, pp. 32, 33, 185-191), and Leuckart (1894, pp. 448-464).

Hosts.—Cattle, sheep, deer, and other animals. (See pp. 137-143.)

Geographical distribution.—Europe, Africa (Egypt), Asia, Australia, Canada, and probably elsewhere.

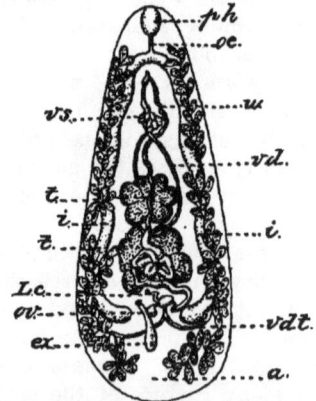

Fɪɢ. 50.—Dorsal view of a Conical Amphistome, showing the anatomical characters: *a*, position acetabulum; *ex*, terminal vesicle of excretory system; *i*, intestinal caeca; *Lc*, Laurer's canal; *oe*, oesophagus; *ov*, ovary; *ph*, pharynx; *t*, testicles; *u*, uterus; *vd*, vas deferens; *vdt*, vitello duct; *vs*, vesicula seminalis. ×5. (After Otto, 1896, p. 100, fig. 4.)

Life history.—Sonsino found in an Egyptian snail a larval parasite (*Cercaria pigmentata*) which, according to some authors, represents the larval stage of this amphistome. The life cycle has recently been experimentally demonstrated by Looss (1896), who describes it as follows.

The eggs escape from the host with the faeces. After a time, evidently varying with the temperature (twelve to fourteen days at 22° C.), a ciliated embryo is formed (fig. 51). This embryo (*miracidium*) escapes from the eggshell only when exposed to light and in case the water is not below 15° C. Swimming around in the water it enters certain snails (*Physa alexandrina* Bourg. and *P. micropleura* Bourg.) establishing itself in the visceral cavity. Here it develops into a *sporocyst* (fig. 52) which, when about fifteen days old, measures 0.7 mm. long by 0.15 mm. broad; a generation of *rediae* (fig. 53) develops in the sporocyst; the rediae escape from the latter in about fifteen days; a second generation of rediae (fig. 54) forms within the first rediae, escaping by the birth opening; a third generation of rediae may develop within the second. The *cercariae* (fig. 55) form in the rediae and are born at an early stage of development; when fully developed these cercariae escape from the snail and swim around in the water. The entire cycle to this point is evidently completed in less than two months. The cercaria (*Cercaria pigmentata*) is oval, 0.5 mm. long by 0.33 mm. broad with a tail about 0.9 mm. long; body opaque, due

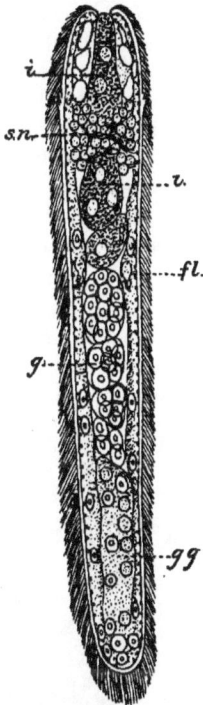

to pigment and to certain subtegumentary cells; oral sucker spherical, 45 μ in diameter; acetabulum .90 μ in diameter; two eye-spots present. The cercariae encyst themselves on plants and various other objects and evidently gain access to the final host (cattle, sheep, etc.) through the drinking water.

FIG. 51.—Dorsal view of the free embryo (miracidium) of the Conical Amphistome (*Amphistoma cervi*) about to enter the intermediate host: *fl*, end portion of excretory system; *g*, germ cells; *gg*, matrix of germ cells; *i*, rudimentary intestine; *sn*, nervous system. ×285. (After Looss, 1896, Pl. XII, fig. 125.)

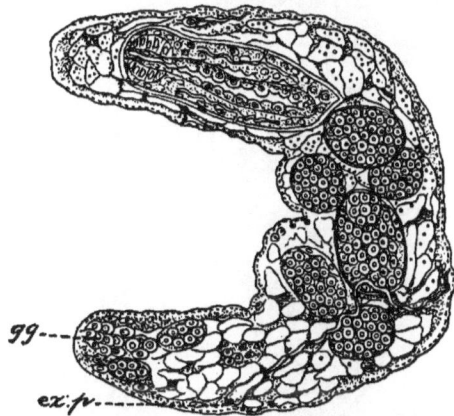

FIG. 52.—Sporocyst of the Conical Amphistome resulting from the transformation and development of the embryo, age about 15 days: *ex. p*, excretory pore. *gg*, matrix of germ cells. The large balls of cells represent developing rediae of the next generation. ×170. (After Looss, 1896, Pl. XII, fig. 126.)

The Conical Fluke seems to have a very wide distribution, being recorded in Europe, Asia, Africa, North America, and South America. As yet it has not been recorded in the United States, but specimens collected in Canada have been sent to us by Professor Wright, and we may expect to find the same worms any day in the United States.

There is considerable difference of opinion among authors as to whether these parasites are injurious to the animals in which they occur. While some writers state that they are absolutely harmless, others claim that they cause an irritation in the stomach, and that cattle which are heavily infested with them gradually emaciate. According to an Australian paper, the parasites cause a considerable number of deaths among the cattle of the coast districts; they occur in great numbers and injure cows more than steers or oxen. Attaching themselves to the mucous membrane of the stomach, by means of their suckers they raise the epithelium in form of papillae. The treatment is the same as for adult tapeworms (see p. 133).

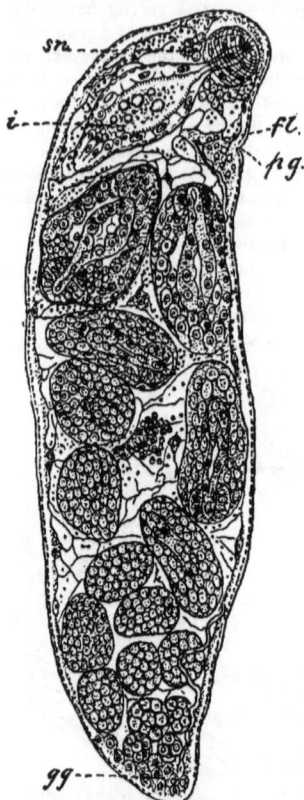

ABATTOIR INSPECTION.

The amphistomes of cattle are of no importance in meat inspection, as they are not transmissible to man in any stage of their development. In fact, according to Schweinfurth, these parasites are collected by the natives of Africa and eaten raw.

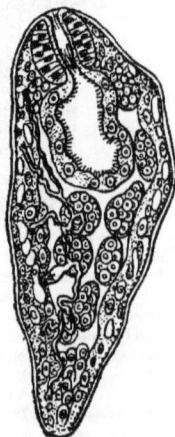

FIG. 53.—Adult redia of the Conical Amphistome (*Amphistoma cervi*) of the first generation, thirty-nine days after the infection of the intermediate host with embryos: *fl*, end portion of excretory system; *gg*, matrix of germ cells; *i*, rudimentary intestine; *pg*, birth opening; *sn*, nervous system. (After Looss, 1896, Pl. XII, fig. 129.)

Several other amphistomes are found in various allied ruminants used for food in certain countries, and although these parasites have not yet made their appearance in this country, we can not tell what moment we shall find them introduced, perhaps with animals imported for menageries. Should they be introduced in this manner and find the conditions necessary to the development of their larval stages, they would, in all probability, develop in our American cattle. As this day has not yet come, they will simply be mentioned here by name and figured. For anatomical characters, compare the figures with the key on page 21.

FIG. 54.—Young redia of the Conical Amphistome of the second generation in which the cercariae develop. ×170. (After Looss, 1896, Pl. XII, fig. 130.)

10. ¹ *Amphistoma explanatum* Creplin is described from the liver and gall bladder of the zebu; von Linstow (1878, p. 49) cites it as a parasite of cattle; Railliet cites it from the Indian buffalo and the zebu.

11. ²*Amphistoma bothriophorum* Braun (fig. 56) also occurs in the stomach of the zebu.

12. *Amphistoma tuberculatum* was reported by Cobbold (1875, p. 819) from the intestine of the Indian oxen; but no description of the parasite has ever been given, so that the form may be ignored.

13. ³*Gastrothylax crumenifer* Creplin. This parasite (figs. 57-62) is said to occur in most of the bovine animals (the zebu) killed at Son-Tay; its natural habitat is the stomach, and when present in large numbers they irritate the mucous lining and lead to an extreme emaciation of their host. The same parasite was once found at Leipsic, Germany, in a cross between the zebu, gayal, and yak; and von Linstow (1878, p. 49) cites it among the parasites of cattle.

14. ⁴*Gastrothylax Cobboldii* Poirier (fig. 63) was described from the stomach of the gayal from Java.

15. ⁴*Gastrothylax elongatum* Poirier (fig. 64) was described from the stomach of the gayal from Java, and Railliet (1893, p. 379) reports that it has been found in Paris in the stomach of a zebu.

16. ⁵*Gastrothylax gregarius* Looss (figs. 65 and 66) is found in enormous numbers in the rumen of nearly all the Indian buffaloes slaughtered in Alexandria, Egypt. In one buffalo Looss counted 1,758 specimens on a portion of the mucosa as large as a hand. They were often found associated with *Amphistoma cervi*.

17. ⁴*Homalogaster paloniae* Poirier (fig. 67) is found in the caecum of the gayal in Java.

18. ⁶*Homalogaster Poirieri* Giard & Billet is found in the large intestine of Tonkin cattle (=? the zebu). It fixes itself by means of the acetabulum to the mucosa and is sometimes present in large numbers.

FIG. 56.—*Amphistoma bothriophorum:* a, position of acetabulum; *ex*, terminal vesicle of excretory system; *i*, intestinal caeca; *Lc*, Laurer's canal; *oe*, oesophagus; *ov*, ovary; *ph*, pharynx; *t*, testicles; *u*, uterus; *vd*, vas deferens. ×5. (After Otto, 1896, p. 102, fig. 5.)

FIG. 55.—Mature cercaria of the Conical Amphistome (*Amphistoma cervi*), the stage which gains access to cattle and sheep. ×75. (After Looss, 1896, Pl. XII, fig. 133.) See p. 65.

---

¹ For original description, see Creplin, 1847, pp. 34 and 35.

² For technical discussion, see Otto, 1896, pp. 101-105.

³ *Amphistoma crumeniferum* Creplin, 1847; *Gastrothylax crumeniferum* (Creplin) Poirier, 1883; *G. crumenifer* (Creplin) Otto, 1896. For bibliography and technical discussion, see Otto, 1896, pp. 94-97.

⁴ For technical discussion, see Poirier, 1883, pp. 73-80.

⁵ For technical discussion, see Looss, 1896, pp. 5-13, 170-177.

⁶ For original description, see Giard & Billet, 1892, pp. 614 and 615.

TAPEWORMS, OR CESTODES (Order Cestoda).

[Segmented Tapeworms (suborder *Tomiosoma*). Tapeworms without four re-
tractile probosces (tribe *Atrypanorhyncha*). Tapeworms with four suckers (subtribe
Tetrassichiona).]

Family TAENIIDAE.

In cattle and sheep we find both of the stages of tapeworms mentioned on page 21, namely:

FIG. 57.—Enlarged dorsal
view of *Gastrothylax
crumenifer*. (After
Creplin, 1847, Pl. II, fig.
1.) See p. 67.

Larval forms (cystic worms, bladder worms, hydatids) which live in the muscles or parenchymatous
organs, but not in the intestine. They render the
meat unfit for food since they are transmissible (according to the species) to man and dogs; and—
Adult worms (tapeworms, stro-
bilae) which occur in the intestines of sheep and cattle (rarely
in the ducts of the liver of sheep)
and are not transmissible to carnivorous animals.

Hogs on the other hand appear to be infested
only with larval tapeworms, although three isolated
cases of adult tapeworms have been recorded for
them. These three cases may have been accidental
occurrences, the hogs having possibly become accidentally infested with worms which
normally live in other animals.

Tapeworms of the family Taeniidae pos-
sess the following characters: The anterior extremity is represented by a more or less knob-like portion known as the *head;* this
is followed by an unsegmented portion, the *neck;* head and neck
together form the *scolex;* this in turn by the *segments*, or *proglottids.*

The head is provided with four cup-shaped suckers, which are
never provided with hooks in any form known in cattle, sheep,
or hogs, but are armed with numerous hook-
lets in some of the forms found in certain other
animals (man, rabbits, birds). The apex of
the head is provided with a muscular body,
which develops into different forms in the various subfamilies.
It may form a *rostellum*, which may be unarmed (*Taenia saginata*)
or armed (*Taenia solium*). In the larval forms discussed in this
paper (*Taeniinae*) the rostellum protrudes at the center of the
apex, but in some other forms (*Dipylidiinae*) it may retract into
a rostellum sac. In the adult tapeworms (*Anoplocephalinae*) of
cattle, sheep, etc., the muscular body is composed of stellate fibers which move
the suckers, but these fibers do not appear to form a true rostellum.

The *neck* is very simple in structure, containing each side two longitudinal
canals and a longitudinal nerve trunk. At the posterior portion of the neck, segments form by transverse division.

The *segments* increase in size, gradually becoming larger the farther they are from
the head; reaching a maximum breadth, they decrease in width, and then increase
in length more rapidly. The anterior segments are the youngest, the posterior seg-

FIG. 58.—Enlarged ven-
tral view of *Gastrothy-
lax crumenifer: a.* ace-
tabulum; *vp,* opening
to the ventral pouch.
(After Creplin, 1847. Pl.
II, fig. 2.) See p. 67.

FIG. 59.—Enlarged
view of anterior
extremity of *Gas-
trothylax crumeni-
fer: m,* mouth; *vp,*
opening to ventral
pouch. (After
Creplin, 1847. Pl.
II, fig. 4.) See p. 67.

FIG. 60.—Enlarged
view of posterior
extremity of *Gas-
trothylax crumeni-
fer*. See p. 67.

ments the oldest. Many zoologists look upon the entire tapeworm as a colony of animals, each separate segment representing a single individual, and all segments being descended from a single animal represented by the head and neck.

Owing to their parasitic life, tapeworms are very degraded in their structure. The *digestive tract* is entirely absent, the worms taking their nourishment by osmosis through their entire surface. The *nervous system* is composed of nerve centers (ganglia), situated in the head, and two large lateral nerves, one of which extends on each side of the worm from the head to the posterior end of the strobila; in some cases, at least, the lateral nerves are connected by two transverse nerves at the distal end of each segment. The *excretory system* consists of two dorsal and two ventral longitudinal lateral canals, which are connected in various ways in the head; the *ventral canals* are connected by *transverse canals* at the posterior border of each segment. The *genital organs* form by far the most important organ system in the animal. In the first place, the entire genital system is repeated, so that each segment as it arrives at a given age possesses its own genital organs, independent of the organs of the other segments. Again, every segment is hermaphroditic, containing both male and female organs, and in some genera the segments are doubly hermaphroditic, containing double sets of male and of female organs.

The *male organs* consist of a cirrus (penis), a cirrus pouch, a vas deferens, and numerous testicles. The *female organs* consist of a vulva, a vagina, an ovary, a vitellogene gland, a shell gland, oviducts, and a uterus. Each segment possesses one or two genital pores, the cirrus and the vulva of any given set of organs (except in the genus *Amabilia*, according to Diamare) opening at the same pore. In some species, so-called *interproglottidal glands* of unknown function are found between the segments.

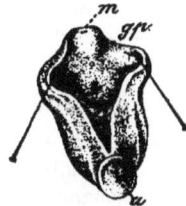

FIG. 61.—Enlarged view of *Gastrothylax crumenifer*, with ventral pouch open: *a*, acetabulum; *gp*, genital pore; *m*, mouth. (After Creplin, 1847, Pl. II, fig. 5.) See p. 67.

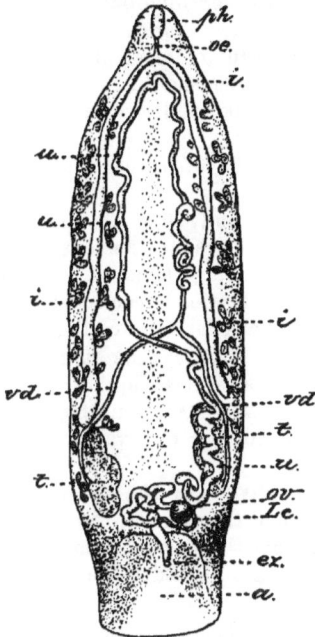

FIG. 62.—Dorsal view of *Gastrothylax crumenifer*, magnified to show the anatomical characters: *a*, acetabulum; *ex*, terminal vesicle of excretory system; *i*, intestinal caeca; *Lc*, Laurer's canal; *oe*, oesophagus; *ov*, ovary; *ph*, pharynx; *t*, testicles; *u*, uterus; *vd*, vas deferens. ×5. (After Otto, 1896, p. 96, fig. 3.) See p. 67.

FIG. 63.—*Gastrothylax Cobboldii*, lateral view: *a*, acetabulum; *i*, intestine; *m*, mouth; *vp*, opening to ventral pouch. (After Poirier, 1883, Pl. II, fig. 3*b*. Taken from Braun, Vermes, Pl. XVIII, fig. 2.) See p. 67.

Life history.—Tapeworms pass through three stages of development, known as the *oncosphere* (or embryo), the larva (a bladder worm known as a *cysticercus*, a *coenurus*, an *echinococcus*, or a *cysticercoid*), and an adult form known as the *strobila*. A change of host is necessary for their development; the host in which the oncosphere develops into the larva is known as the *intermediate host*,

while the animal which harbors the adult form is known as the *final host*. The life history of *Taenia saginata* given on page 72 may be taken as typical for the family.

All of the larval cestodes of cattle, sheep, and swine belong to the subfamily *Taeniinae*, while all the adult forms found in these hosts are classified in the subfamily *Anoplocephalinae*.

Hard-shell Tapeworms (Cestodes of the Subfamily *Taeniinae*).

The Hard-shell Tapeworms, so called because of the thick striated eggshell (embryophore), are found as adults in the intestines of meat-eating mammals, while their larval stage is found in the muscles or parenchymatous organs of herbivorous and omnivorous animals. These larval forms are very important from the standpoint of meat inspection, and organs which harbor them should be excluded from the market or should be rendered wholesome before being placed on sale.

The larval forms may be of three kinds, as follows:

(1) *Cysticercus* (figs. 68 and 76).—This is the most simple form. The parasite consists of a cyst, which is invaginated at a given point. There is normally only one invagination to each cyst, and at the base of the invagination is situated the head of the future tapeworm. Besides the invagination, the cyst contains more or less liquid.

(2) *Coenurus* (figs. 99 and 100).—In this case there is a considerable number of invaginations, each containing a head.

(3) *Echinococcus* (fig. 105).—In the third type there is no invagination of the cyst wall, but brood capsules are formed from the parenchyma of the cyst and several heads are formed in each brood capsule.

FIG. 64.—*Gastrothylax elongatum:* g, ganglion; i, intestinal caeca; ph, pharyngeal bulb; t, testicle; u, uterus; vd, vas deferens (ductus ejaculatorius); ve, vasa efferentia; vg, vitellogene glands. (After Poirier, 1883, Pl. II, fig. 2b. Taken from Braun, Vermes, Pl. XVIII, fig. 7.) See p. 67.

Attempts have been made to subdivide the *Taeniinae* into genera and subgenera, the genus *Taenia* Linnaeus being retained for the forms which possess a *Cysticercus* or a *Coenurus* as larval form, while *Echinococcifer* Weinland, 1861, has been proposed as the generic name for *Taenia echinococcus*. This generic division has not been accepted by the majority of helminthologists, most workers preferring to recognize only one genus, *Taenia*, but many authors admitting three subgenera, corresponding to the three types of larvae.

HARD-SHELL TAPEWORMS (Genus Taenia).

The following species of this genus must be considered in this report:

Adult.		Larva.	
Name.	Host.	Name.	Host.
Taenia saginata	Man	*Cysticercus bovis*	Cattle.
Taenia solium	Man	*Cysticercus cellulosae*	Swine and man.
Taenia marginata	Dogs	*Cysticercus tenuicollis*	Cattle, sheep, and swine.
Taenia coenurus	Dogs	*Coenurus cerebralis*	Cattle and sheep.
Taenia echinococcus	Dogs	*Echinococcus polymorphus*	Cattle, sheep, swine, man, etc.

19. Beef Measles (*Cysticercus bovis*) of Cattle, and its adult stage, The Unarmed, or Beef Measle, Tapeworm (*Taenia saginata*) of Man.

[Figs. 68–74.]

LARVAL STAGE (*Cysticercus bovis*).

For anatomical characters, compare fig. 68 with key, p. 21.

SYNONYMY.—*Cysticercus Taeniae saginatae* Leuckart, 18—(?); *C. bovis* Cobbold,

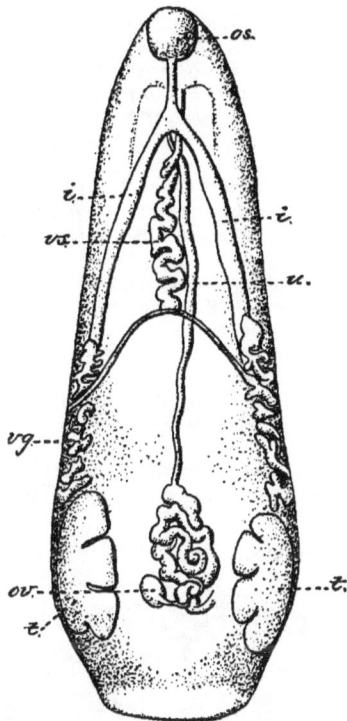

FIG. 65.—Dorsal view of *Gastrothylax gregarius*: *i*, intestinal caeca; *os*, oral sucker; *ov*, ovary; *t*, testicles; *u*, uterus; *vg*, vitellogene glands; *vs*, vesicula seminalis. ×9. (After Looss, 1896, fig. 1.) See p. 67.

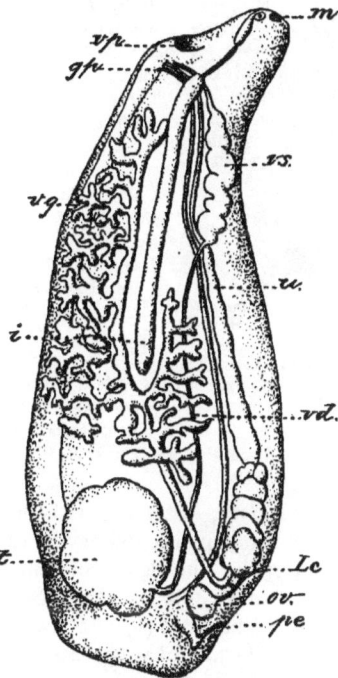

FIG. 66.—Lateral view of *Gastrothylax gregarius*: *gp*, genital pore; *i*, intestinal caeca; *m*, mouth; *Lc*, opening of Laurer's canal; *ov*, ovary; *pe*, excretory pore; *t*, testicle; *u*, uterus; *vd*, vas deferens; *vp*, ventral pouch; *vg*, vitellogene gland; *vs*, vesicula seminalis. ×9. (After Looss, 1896, fig. 2.) See p. 67.

1866; *C. Taeniae mediocanellatae* Knoch, 1868; *C. inermis* of various Germans and others, 18—(?); "*Cysticerkus*" *bovis* of Schneidemühl, 1896.

HOSTS.—Cattle, Rocky Mountain "antelope," llama, and giraffe. (See pp. 137–143.)

ADULT STAGE (*Taenia saginata* (Goeze, 1782)).

For anatomical characters, compare figs. 69–73 with key, p. 84.

SYNONYMY (see also pp. 89–90).—*Taenia solium* Linnaeus, 1758, pro parte; *T. cucurbitina* Pallas, 1781, pro parte; *T. cucurbitina* Art [=var.] *saginata* Goeze, 1782; *T. cucurbitina, grandis, saginata* Goeze, 1782; *T. solitaria* Leske, (1785), pro parte; *Halysis solium* (Linnaeus) Zeder, 1803, pro parte; *Pentastoma coarctata* Virey, 1823; *"T. dentata"* Nicolai, (1830) [nec Batsch, 1786]; *"T. lata"* Pruner, 1847 [nec Linnaeus, 1758]; *Bothriocephalus tropicus* Schmidtmüller, 1847; *T. mediocanellata hominis*, seu *T. mediocanellata* seu *T. zittaviensis* Küchenmeister, 1852; *T. solium* var. *mediocanellata* (Küchenmeister) Diesing, 1854; *Taeniarhynchus mediocanellata* (Küchenmeister) 1858; (?) *Taenia solium* var. *abietina* Weinland, 1858; *T. inermis* Moquin-Tandon, 1860; *T. mediocancellata* (-?-), date (?), see Moquin-Tandon, 1860; *T. tropica* (Schmidtmüller) Moquin-Tandon, 1860; *T. megaloon* Weinland, (1861); *T. (Cystotaenia) mediocanellata* of Leuckart, 1863; *T. saginata* (Goeze, 1782) of Leuckart, 1867; (?) *T. abietina* Weinland of Davaine, 1873; *T. inermis* Laboulbène, 1876; *T. algérien*, Redon, 1883; (?) *T. solium* var. *minor* Guzzardi Asmundo, 1885; *T. algeriensis* Braun, 1894 (= *T. algérien* Redon renamed).

ANOMALIES.—(?) *"Taenia vulgaris"* Werner, 1782 [nec Linnaeus, 1758] = *T. dentata* Batsch, 1786; (?) *T. fenestrata* Chiaje, 1833; *T. capensis* Moquin-Tandon, 1860; *T. lophosoma* Cobbold, 1866; *T. fusa*, *T. continua*, *T. solium fusa* seu *continua* Colin, 1876; *T. mummificata* Guzzardi Asmundo, 1885; *T. nigra* Davaine, 1877; *T. inermis fenestrata* Maggiora, 1891.

PRE-LINNAEAN NAMES.—*Vermis cucurbitinus* composing *Taenia longissima* Plater, 1609; *Lumbricus latus* Movfetus, (1634); *Taenia secunda* Plateri Ernst, 1659; *Lumbricus latus* Tyson, 1683; *Solium* ou *Ténia sans épine* Andry, (1700); *Taenia de la seconde espèce* Andry, 1718; *Taenia sans épine* ou *Taenia de la première espèce* Andry, 1741.

Fig. 67.—*Homalogaster paloniae*, ventral view. (After Poirier, 1883, Pl. II, fig. 1a. Taken from Braun, Vermes, Pl. XVIII, fig. 3.) See p. 67.

BIBLIOGRAPHY.—For bibliography, see Huber (1892). For technical discussion, see Leuckart (1880, pp. 513–616); R. Blanchard (1886, pp. 315–382). HOST.—Man.

Life history.—Starting with the adult tapeworm (fig. 69) in the intestine of man, the life history of the parasite, the knowledge of which we owe to Rudolf Leuckart, is as follows: The eggs (fig. 74) escape from the uterus and are passed with the excreta, or the segments containing eggs break loose from the tapeworm and either wander out of the intestine of their own accord or are passed with the excreta. In either case the eggs become scattered upon the ground or in water, and reach the cattle through their drinking water or with the fodder. When whole segments (generally several together) are passed, these crawl around on the ground or herbage, and cattle by swallowing them

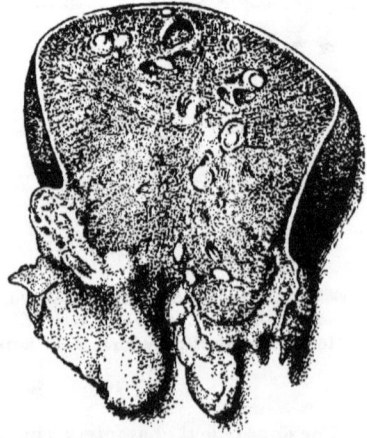

FIG. 68.—Section of a beef tongue heavily infested with beef measles, natural size (original).

may become infected with numerous eggs at the same time. Upon arriving in the stomach, the eggshells are destroyed; the embryo then bores its way through the intestinal walls with the aid of its six minute hooks, and wanders to the muscles where it comes to rest; or if it bores into a blood vessel, it may be carried with the blood to any organ of the body. When the embryo comes to rest it loses its hooks, and, increasing in size, develops into a small round bladder worm. The head of the future tapeworm is then developed in an invagination of the cyst wall, and the complete organism (fig. 68) thus formed is known as a *cysticercus*, or bladder worm. During its development the cyst pushes the tissues of the host aside to make room for itself and an outer cyst is formed around it, made up of connective tissue of the host. The total time consumed in the development of the cysticercus from the embryo is variously estimated from seven to eighteen weeks.

Hertwig states that the larva has completed its development in eighteen

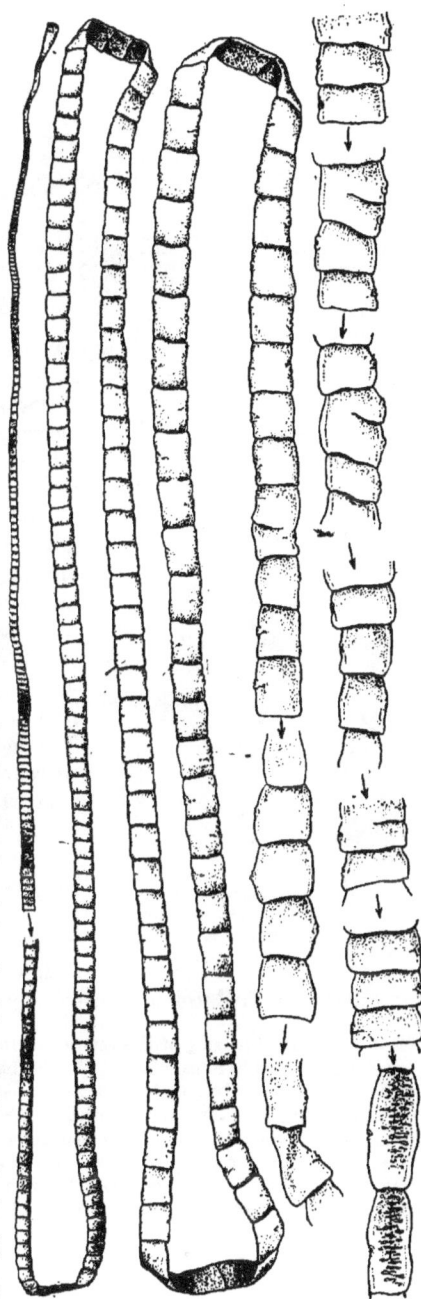

FIG. 69.—Several portions of an adult Beef-measle Tapeworm (*Taenia saginata*) from man, showing the head on the anterior end and the gradual increase in size of the segments, natural size (original).

weeks, and gives the following table to determine the age[1] of the cysticercus:

Age in weeks.	Entire cyst.	Cysticercus without connective tissue cyst.	Scolex.	
			Natural size.	Stretched.
4	4.0 by 3.5 mm	2.25 by 2.25 mm	0.5 by 0.5 mm	0.7 mm. long.
6	4.2 by 3.5 mm	3 by 2.5 mm	1 by 1 mm	1.3 mm. long.
8	4.5 by 3.5 mm	3.25 by 2.75 mm	1.5 by 1 mm	2.9 mm. long.
10	5 by 3.75–4 mm	3.5 by 3.5 mm	1.75 by 1 mm	3.3 mm. long.
12	5.6 by 3.5–4 mm	4 by 4 mm	1.8 by 1 mm	3.5 mm. long.
14	6 by 4.5 mm	5 by 4.5 mm	2 by 1 mm	4 mm. long.
16	6 by 4.5 mm	5 by 4.5 mm	2 by 1 mm	4.25 mm. long.
18	6.25–7 by 4.5 mm	6 by 4 mm	2 by 1.25 mm	5 mm. long.
22	6.5–8 by 4.5 mm	6 by 4.5 mm	2.25 by 1.75 mm	5.5–6.25 mm. long.
28	7.5–9 by 5.5 mm	7 by 5 mm	2.5 by 2 mm	7 mm. long.

The calcareous bodies may appear when the bladder worm is four weeks old; the suckers are fully developed at the eighteenth week. Occasionally bladder worms are found measuring 10 to 12 mm., with a scolex 8 to 9 mm.; these parasites are more than 28 weeks old. If the infested animal is not slaughtered and does not die the cysts will eventually die and degenerate. Thus, in one animal killed 224 days after being infected with tapeworm eggs, the cysts were calcified (Saint-Cyr). The parasites which inhabit the seats of predilection (see p. 78) seem to be the last to die. The degeneration may include (1) the surrounding connective tissue capsule, which becomes opaque and thickened, (2) the bladder cyst of the worm, which turns to a yellowish green, soft, cheesy mass, or (3) both. If, however, the animal is slaughtered before the cysticerci become calcified, and the meat used for food, the cyst around the hydatid is digested upon arriving in the stomach of

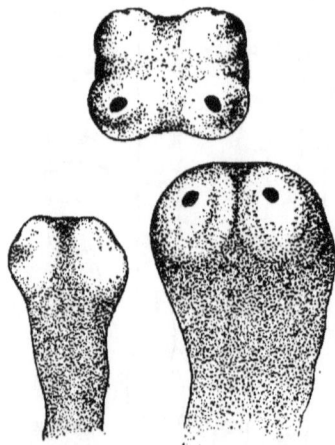

FIG. 70.—Dorsal, apex, and lateral views of the head of Beef-measle Tapeworm (Taenia saginata), showing a depression in the center of the apex. ×17. (Original.)

---

[1] During the proof reading of this report an article by Ostertag (1897, pp. 1–4) has reached us, in which he adds some details of value in judging the age of the bladder worms. His chief results may be summarized as follows:

(1) A steer may become infected with beef measles, and yet recover from the attack without showing upon post-mortem any calcified cysts. (2) C. bovis, 18 days old, is spindleform, and measures 4 mm. long by 2 mm. broad; a differentiation into scolex and bladder is not yet present. (3) Up to 33 days after infection the parasite is surrounded by a cheesy mass, the result of exudation; this afterward disappears. (4) At 25 days old the parasite shows the primordium of the scolex, with faint indication of the suckers. (5) When the parasite is 59 days old the suckers may be seen with the naked eye; calcareous corpuscles are also present. (6) The lumen of the suckers is visible in parasites 73 days old.

man. The hydatid cyst is also digested, the head and neck alone remaining uninjured. The scolex then passes from the stomach into the small intestine, fastens itself to the wall by means of its suckers, and gives rise to segments by transverse division (strobilization) directly back of the head and neck. New segments are formed between the head and the old segments, so that the last segment is always the oldest and the segment nearest the head always the youngest. Segments are formed so rapidly that the worm is full grown at the end of about three months. Perroncito estimates that about 13 to 14 new segments are formed each day, which results in an average increase in the length of the worm at the rate of about 3 cm. per day for the first month and 14 cm. per day for the second month. Genital organs, both male and female, are developed in every segment, embryos are produced, and the life cycle is completed to the point from which we started out.

BEEF MEASLES.

The disease in cattle in the skel cattle.—*Cysticercus boris* has been found in etal muscles, in the heart, the adipose tissue around the kidneys, the subperitoneal connective tissue, the lymphatic glands, and between the convolutions of the brain; cases are also reported of its presence in the lungs and liver.

In some infections of cattle which have been made no symptoms of disease were noticed, but in others quite severe symptoms have been observed. About fifteen to twenty days after infection the animals became feverish, the sickness increasing to the twenty-fifth to sixtieth day, the patients becoming emaciated. Several cases

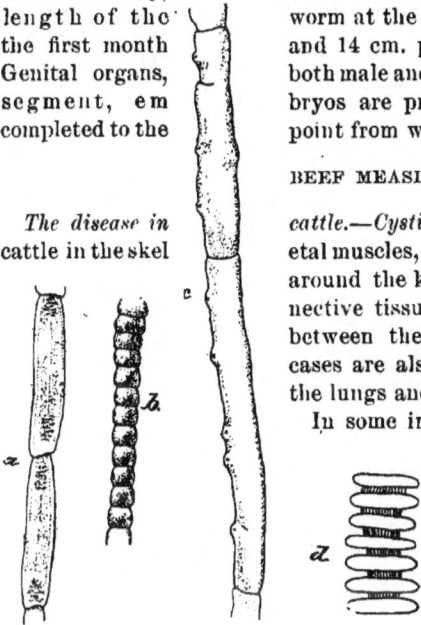

Fig. 71.—Segments from various strobilae of Beef-measle Tapeworm (*Taenia saginata*) showing forms of proglottids which are occasionally found: *a*, elongated segments; *b*, beadlike segments; *c*, a portion of strobila in which the segmentation is not distinct; *d*, moniliform segments (*a* and *b* original; *c* and *d* after R. Blanchard, 1894).

proved fatal, while others recovered and were apparently none the worse for the experiment.

These symptoms have been noticed only in cattle which have been experimented upon and which have received enormous infectious, and it is very generally supposed that an ordinary infection will have little or no effect upon the animals; we can, however, easily imagine that such an infection as Fleming describes, where he found 300 cysticerci in one pound of muscle, will injure the host. When the heart is heavily infected its action must be seriously impeded. As an example of an extreme case we may take the following description of symptoms and post-mortem examination, taken from Zürn (1882, p. 187):

Symptoms.—Four days after feeding segments of *T. saginata* to a healthy three-months-old calf, the patient showed a higher temperature (the normal temperature was 39.2 C.) The calf ate but little on that day, showed an accelerated pulse, swollen belly, staring coat, and upon pressure on the sides showed signs of pain. The next day the animal was more lively, ate a little, and for nine days later did not show any special symptoms except pain on pressure of the abdominal walls, and a slight fever. Nine days after the infection the temperature was 40.7 C., pulse 86, respiration 22; the calf laid down most of the time, lost its appetite almost entirely, and groaned considerably. When driven it showed a stiff gait and evident pain in the side. The fever increased gradually and with it the feebleness and low spiritedness of the calf, which now retained a recumbent position most of the time, being scarcely able to rise without aid, and eating only mash with ground corn. Diarrhea commenced, the temperature fell gradually, and on the twenty-third day the animal died. The temperature had fallen to 38.2 C. During the last few days the calf was unable to rise; in fact, it could scarcely raise its head to lick the mash placed before it. Pulse was reduced by 10 beats. On the last day the heart beats were

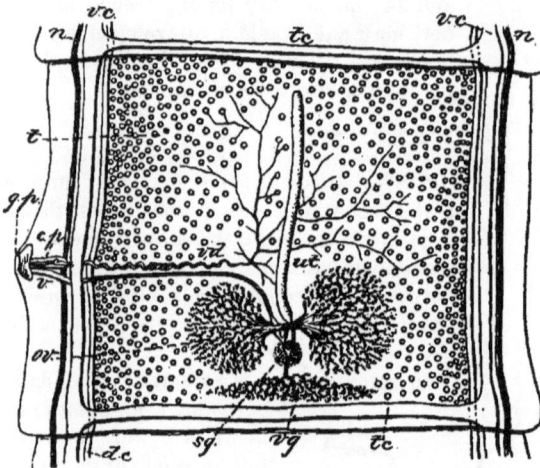

FIG. 72.—Sexually mature segment of Beef-measle Tapeworm (*Taenia saginata*): *cp*, cirrus pouch, with cirrus; *dc*, dorsal canal; *gp*, genital pore; *n*, lateral longitudinal nerves; *ov*, ovary; *sg*, shell gland; *t*, testicles; *ut*, median uterine stem, enlarged (in part after Leuckart); *v*, vagina; *vc*, ventral canal, connected by transverse canal, *tc*; *vd*, vas deferens; *vg*, vitellogene gland.

very much slower, yet firm, and could be plainly felt. Several days before death the breathing was labored and on the last day there was extreme dyspnoea. * * *

Post-mortem.—Body cavities contained reddish serous exudate. Subdermal connective tissue was oedematous. Muscles were redder than usual, in some places very dark red. In the heart muscles were innumerable (many thousand) round tubercle-like bodies, 1.5 to 3 mm. long, 1.2 to 5 mm. wide, yellowish-white in color. Young cysticerci lay embedded in these smeary chalky cysts. Some of these cysticerci were round, but the majority were bottle-shaped and contained round cells and fat globules, and were inclosed by a membrane.

The bottle-shaped cysticerci measured 0.557 mm. long while their greatest diameter was 0.326 mm. Cysts were also found in all the muscles, especially in the muscles of mastication, dorsal muscles of the neck, etc., and finally, though not many, in the diaphragm, and outer and inner diagonal abdominal muscles.

Acute cestode tuberculosis is a name which is sometimes applied to designate a heavy infection with cysticerci.

Failing to diagnose the presence of the parasites by symptoms exhibited by the cattle, we have recourse to other methods which have occasionally proved of use:

(1) By examining the under side of the tongue it is occasionally possible to find small lumps about the size of a pea or bean which can be moved slightly with the fingers, and which in many cases represent cysticerci. It is rare that this method leads to any practical results, and no confidence should be placed in it.

(2) Several authors suggest cutting out a portion of muscle—one of the neck muscles, for instance—and examining it for the cysts. Although the diagnosis may sometimes be made by operating in this manner, we can hardly see how it can be of any practical value, since no treatment, except good nourishment, which cattle should always have, can at present be suggested for animals infested with these parasites. Moreover, extirpation of a muscle should be practiced only by professionals, and a negative diagnosis in this case is of no value.

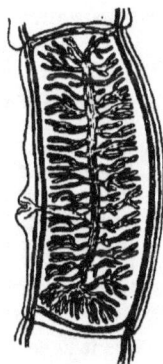

(3) The only positive diagnosis is post-mortem examination, and this, for the comfort of man, as will be shown, should be made on all slaughtered cattle. An examination of the internal and external muscles of the jaws, the tongue and neck, as well as the heart and muscles seen from the body cavity, will generally suffice to determine whether the cysticerci are present or not.

Treatment.—There is no medical treatment to be suggested. *Prevention*, however, is extremely simple. We have seen (p. 72) that cattle obtain the eggs directly or indirectly from human excrements; hence persons who have this tapeworm should not void their excrements in fields or barns where they can contaminate the fodder or water used for cattle. If this plan is followed, not only will the spread of the parasite among cattle be prevented, but also the spread of this species of tapeworm among man, since the latter, as has already been stated, becomes infected by eating meat containing the larval stage. It lies entirely within the power of the inhabitants in stock-raising districts to prevent the infection of their cattle with this parasite.

Fig. 73.—Gravid segment of Beef-measle Tapeworm (*Taenia saginata*), showing lateral branches of the uterus, enlarged (original).

ABATTOIR INSPECTION.[1]

As beef measles, when swallowed by man, gives rise to an adult tapeworm, the question of using meat infested with this parasite and

<hr>

[1] Since this report was sent to press we have received a very extensive article upon abattoir inspection for beef measles, written by Rasmussen (1897), of Kopenhagen. Persons who wish to inform themselves upon this subject more in detail are referred to the articles by Rasmussen (1897) and Friis (1897).

the question of abattoir inspection naturally arise. Although this species of tapeworm does not contribute to the comfort of man, it can not
be looked upon as a dangerous parasite. It may bring about digestive
troubles, but will not per se result in the death of the patient; and
with proper treatment it can be gotten rid of, although sometimes
with difficulty. Nevertheless, its presence in man should, of course, be
prevented when possible, and this can be done by very simple means,
namely, (1) by an inspection of cattle at the slaughterhouses to determine the presence or absence of the larval stage, and (2) by submitting
infested meat to processes which kill the parasites.

(1) *Position of the parasites.*—Inspectors should examine very thoroughly the muscles of all cattle slaughtered, especially the inner and
outer muscles of mastication and the heart; cuts should be made into
the muscles of the jaws parallel to the bones.

The following table, taken from Ostertag, giving the result of the
meat inspection in Berlin for 1889–90, is exceedingly instructive, as it
shows the general distribution of the parasites in the various muscles:

	Cases.
(1) In muscles of the jaws	316
(2) In muscles of the jaws and in the heart	39
(3) In muscles of the jaws and in the tongue	4
(4) In muscles of the jaws and in the neck	1
(5) In muscles of the neck	1
(6) In muscles of the neck and in the tongue	1
(7) In muscles of the tongue	2
(8) In muscles of the tongue and heart	2
(9) In muscles of the tongue and muscles of breast	1
(10) General infection	22

Thus it is seen that in these examinations the muscles of the jaws
were infested 360 times, while the other organs were infested but 55
times (in this computation the 22 cases of general infection are omitted);
in other words, in about seven-eighths of all cases found parasites
were present in the muscles of the jaws. Occasionally, in very heavy
infections, the parasites occur also in the lymphatic glands, the lungs,
the liver, the brain, etc.

The recognition of the fully developed bladder worms is an easy matter for anyone who understands the structure of the parasite; although
their detection in superficial layers is rendered somewhat difficult in
case the surface of the meat becomes dried. In case of doubt the suspected cysticercus may be placed between two fingers and a gradual
pressure exerted upon the cyst. This will cause the protrusion of the
head, upon which the four suckers can be easily distinguished. A simple microscopic preparation of the parasite, made by pressing it between
two pieces of glass, will reveal the presence of the calcareous corpuscles of the parenchyma.

Differential diagnosis.—The only parasites in cattle which would be
likely to be mistaken for beef measles are *Cysticercus tenuicollis* (see

p. 96) and *Echinococcus polymorphus.* The absence of hooks, however, immediately distinguishes the Beef-measle Bladder Worm (*Cysticercus bovis*) from *C. tenuicollis*, which occurs in the serous membranes, etc., but not in the muscles. Young stages of the echinococcus hydatid, which are occasionally found in the muscles, differ from the Beef-measle Bladder Worm in several characters, which render a differential diagnosis comparatively easy, as seen from the following table:

Cysticercus bovis.	*Echinococcus,* p. 113.
One unarmed head present..	Head absent;or numerous armed heads present in brood capsules (p. 116).
Cuticle thin.................	Cuticle thick and laminated (p. 116).
Form oval....................	Form round.

A positive diagnosis of the younger stages of *C. bovis* (i. e., before the head has developed), or of degenerated specimens, is sometimes more difficult; the younger stages and the totally degenerated specimens will not develop further if eaten; the specimens which are only partially degenerated may, however, still retain enough vitality to develop into adult tapeworms. The oval or pyriform body gives a probable diagnosis for the younger stages, while the presence of calcareous corpuscles (seen only with the microscope) furnishes a method of diagnosis for the degenerated forms. Even in completely degenerated cysts the calcareous bodies may be discovered; these should, however, not be mistaken for fat globules which are more strongly refractive, possess a broader and darker edge, and do not change on addition of acetic acid. It is more difficult to distinguish the calcareous corpuscles from certain crystals of calcium carbonate; the latter lie in clumps and overlap each other, and upon being treated with mineral acids (as weak hydrochloric acid) completely disappear, while when the calcareous corpuscles are treated with acids their organic base retains the original form.

Rissling gives the following method for determining the presence of cysticerci in chopped meat and sausage, but its application does not seem very practicable, for this country at least. It appears to us much better to inspect meat for measles before it is cut up.

Prepare 1 to 4 liters of a solution of caustic soda or caustic potash having a specific gravity of 1.15; place this, together with the teased or chopped meat, in a funnel-shaped dish, stir well and allow to stand. The worms will then sink to the bottom while the rest of the material will float.

Schmidt-Mülheim's method consists in artificially digesting the meat at 40° C. After several hours the bladder portion of the cysticerci will be more or less destroyed, but the heads will not be affected; they sink to the bottom of the vessel and may be recognized as small white bodies. In armed cysticerci (*C. cellulosae*, p. 89, etc.) the hooks will be found.

Frequency of Cysticercus bovis *in cattle.*—No exact statistics have been published for this country. The proportions of infected cattle slaughtered in Prussia for 1892 were as follows:

Regierungsbezirk.	Proportion.	Regierungsbezirk.	Proportion.
Stralsund	1:51	Kassel	1:1,053
Oppeln	1:229	Bromberg	
Hannover	} 1:350	Breslau	} 1:2,500
Lüneburg		Arnsberg	
Danzig	1:426	Koblenz	
Berlin	1:610	Stettin	
Marienwerder		Posen	} 1:3,500
Frankfurt	} 1:775	Königsberg	1:6,659
Merseburg		Liegnitz	1:20,000
Schleswig	1:915	Düsseldorf	1:25,586
Potsdam	} 1:1,025	Wiesbaden	1:34,182
Hildesheim			

The average proportion for the first 20 Regierungsbezirke mentioned was 1:1,631. These statistics appear rather low when we notice the following figures for the Berlin abattoir:

Year.	Cattle slaughtered.	Cattle infected.	Proportion.
1883–88		2	
1888–89	141,814	113	1:1,255
1889–90	154,218	390	1:395
1890–91	124,593	263	1:474
1891–92	136,368	254	1:541
1892–93	142,874	214	1:672

This apparent increase in proportion from 1:1,255 in 1888–89 to 1:672 in 1892–93 is due to the more thorough inspection following Hertwig's discovery of the seats of predilection of the parasite, rather than an actual increase in the number of animals infected.

Influence of age and sex of the host.—According to certain European statistics about 50 per cent of the cases of infection are found in animals 2 years old; about 20 per cent of the cases in animals 3 years old, and about 4.5 per cent in animals of 1, 4, 5, 6, 7, and 8 years old, respectively. Beef measles are also said to be more common in male animals than in female animals.

Influence of season.—According to Rasmussen, bladder worms are more common in late summer and early fall than at other times of the year. From statistics he gives for Copenhagen it appears that for the years 1890–96, the total number of cases found and their proportion to the entire number of animals slaughtered were as follows:

Months.	Cases.	Per cent.	Months.	Cases.	Per cent.
January	39	0.17	July	13	0.10
February	38	.19	August	52	.31
March	43	.20	September	69	.32
April	20	.10	October	81	.35
May	24	.12	November	54	.24
June	13	.09	December	38	.18

Disposition of measly beef.—Measly beef should be condemned to the tank as unfit for food when the infection is general, or when the inva sion by the parasites has caused a watery and " flabby" condition of the meat. In case of light infection the meat can be used for food after the cysticerci have been rendered harmless, but even in these cases it is well to cut away the most heavily infested portions. Cases of so-called "single infection" should be treated the same as cases of light infection, for although it may unquestionably happen that an animal is infected with but one bladder worm, still the finding of only one parasite is no proof that other parasites are not present; further-more, in a number of cases of alleged " single infection," later and more thorough examination has revealed further worms.

In case of infection with only very young parasites, in which the suckers are not fully developed, the meat may safely be passed and allowed to go on the market without restriction.

In case of infection with fully developed live bladder worms, the meat should be subjected to some safeguarding method before being placed on the block, or it should be sold under declaration of its exact character.

Opinion differs as to the method which should be followed in case of infection with degenerated bladder worms. It is maintained by some that this meat should be allowed on the market without restrictions. The finding of degenerated cysts, or bladders, however, is no proof that all the parasites are dead, for

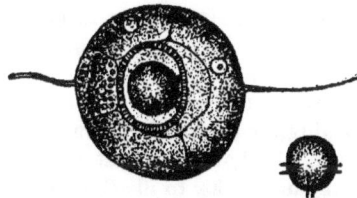

FIG. 74.—Egg of Beef-measle Tapeworm (*Taenia saginata*), with thick eggshell (embryophore), containing the six-hooked embryo (oncosphere), enlarged. (After Leuckart.)

not only are cases more or less frequently found in which both live and degenerated bladders are present, but even if the cyst, or bladder, is degenerated the head may, in some cases, still retain its vitality. It is accordingly safer to treat carcasses with degenerated cysts in the same manner as carcasses with live cysts; and should any exception to this be made, such exception should be limited to cases in which the degenerated parasites are found in the muscles of the jaws.

The cysticerci may be killed by cooking, by salting or corning, or by cold storage.

Cooking.—This is the surest method of killing the parasites, but it is open to the serious practical objection that, according to estimates, cooking in a steam sterilizing apparatus results in a shrinkage of from 33 to 50 per cent, and this heavy loss will undoubtedly be a drawback to its general use.

Perroncito found that below 30° C. the movements of the worms are very slight, or practically nil; from 36° to 38° C. the movements are livelier; at higher temperature they diminish, ceasing at 44° C.; they

"die sometimes at 44° C., now and then at 45° C., and always at 46° C."
He " therefore concluded that they could in no case survive 47° C.
and 48° C. [= 116.6° to 118.4° F.] when they are maintained at this
temperature for at least five minutes. Lewis found a somewhat higher
temperature necessary in order to kill the worms. He states—

(1) That exposure to a temperature of 120° F. for five minutes will not destroy life
in cysticerci, but they may continue to manifest indications of life for at least two
or three days after such exposure; (2) that exposure to a temperature of 125° F. for
five minutes does not kill them, but (3) after being subjected to a temperature of
130° F. for five minutes they may be considered to have perished. After exposure
to this and higher temperatures, in no instance have I been able to satisfy myself
that the slightest movements took place in their substance when examined even
under a high power. At least, it may be confidently asserted that after exposure
for five minutes to a temperature of 135° to 140° F. life in these parasites may be
considered extinct.

Pillizzari found that cysticerci died at a temperature of 60° C.
(= 140° F.), while according to Hertwig 52° C. (= 125.6° F.) reduces
the bladder worm to a smeary, soft condition, so that it can be easily
flattened out between two pieces of glass. It is important to recall,
however, that in cooking large pieces of meat the temperature of the
inner portion does not rise as rapidly as that of the outer portion As
an index to the duration of cooking required in order to guaranty that
all the bladder worms are killed, Ostertag gives the rule of two hours'
cooking for pieces of varying length, but not over 12 cm. (= 6 inches)
thick. Probably the best criterion in forming a judgment is the color
of the meat; 60° to 70° C. (= 140° to 158° F.) causes a reduction of the
haemoglobin, and this results in giving a gray color to beef and a white
color to pork; when slices of cooked beef (or pork) assume this gray
(or white) color, it can safely be assumed that all the cysticerci have
been killed.

In 1894 and 1895 Berlin, Prussia, cooked 342 insured beef-measly
carcasses, representing insurance policies to the value of 57,223.30
marks (about $13,619.15). During the same period there were con-
demned 221 uninsured beef-measly carcasses, valued, on the same
basis, at 36,977.60 marks (about $8,800.67). The raw beef was sold to
the parties having the cooking in charge at 20 pfennige (about 5 cents)
per pound, and after being cooked was sold to the public at 30 to 35
pfennige (about 7½ to 8¾ cents) per pound. Cooked measly pork was
sold at 40 pfennige (about 10 cents) per pound.

Salting.—Salt solution kills the bladder worms in twenty-four hours,
the parasites becoming shriveled. Here, again, it must be remembered that it takes some time for the salt to reach the deeper layers.
• It is probable that this method will to some extent supersede the
cooking, since the shrinkage by salting is estimated at only 6.6 per
cent. Two carcasses of 500 pounds each, treated at Kiel by different
methods, form an excellent comparison. The cooked carcass gave 300
pounds of beef which sold at 30 pfennige (about 7½ cents per pound; in

all, 90 marks (about $21.42); the salted carcass gave 460 pounds of beef which sold at 40 pfennige (about 10 cents) per pound; in all, 184 marks (about $43.79).

In Saxony measly beef has sold as high as 1 mark (about 24 cents) per pound, while in some parts of Germany it has been sold from 2½ to 10 cents per pound. In a few instances the prejudice against this meat was so great that it could not be sold at all. In general, however, the salted measly beef is easier to dispose of than the cooked measly beef.

Ostertag gives as rule for salting the following: Cut the meat into strips of any given length, but not over 6 cm. (3 inches) in thickness, and place for two weeks in a brine composed of 1,000 parts of water, 250 parts of salt, 20 parts of sugar, and 2½ parts of saltpeter. As a practical test to determine whether the salting is thorough, and the parasites are dead, he suggests the use of a 1 per cent solution of nitrate of silver. If applied to the surface of lightly cured meat this solution will produce no change in the appearance; if the meat is fully cured, a momentary milky opacity will result, owing to the formation of chloride of silver. To use the test, wash the meat thoroughly in water, then wipe it with a cloth, and make a quick incision through the middle of the piece to be tested; apply a few drops of the nitrate of silver to the cut surface.

Cold storage.—Perroncito maintains that *Cysticercus bovis* dies fourteen days after its host has been slaughtered. More recent investigations by Ostertag, Zschokke, Glage, and others have shown that two weeks form too short a limit, but that none of the worms can survive three weeks; beef-measly meat which has been in cold storage for three weeks may therefore be looked upon as harmless.

In view of these recent investigations, I can see no reason why light cases of beef measles (but not pork measles, see p. 94), which have remained three weeks in cold storage, should not be passed as first-class meat and allowed on the open market without further restrictions. During certain seasons of the year, however, there is a practical objection to this method of safeguarding which has been thus far overlooked (except by Friis). Experience has shown that meat which has been in cold storage for this length of time during summer will spoil much more rapidly when taken out of the cooler than meat which has been placed in the ice box only long enough to cool and "firm."

THE ADULT TAPEWORM IN MAN AND METHODS OF PREVENTING THE INFECTION OF CATTLE.

Taenia saginata, or the large Unarmed Tapeworm, is the most common of the ten species of tapeworms found in the intestine of man. A form with which it has frequently been confounded is a tapeworm (*T. solium*) of about half the size (2 to 3.5 m.), acquired by eating pork infected with larvae (*Cysticercus cellulosae*), which are very similar to those found in the cattle, but are somewhat larger and possess a double crown of hooks on the head.

The Unarmed Tapeworm of man is almost cosmopolitan, and is especially common in Africa and Asia. Many of the published statistics of the relative frequency of *T. saginata* and *T. solium* are, however, to be taken with reserve. In some countries the Beef-measle Tapeworm is said to be increasing and the Pork-measle worm to be decreasing in frequency; but in some cases these statements are unquestionably based upon misdeterminations. Physicians too frequently make their determinations upon the external form of the segments—a method which can not be relied upon, even when such determination is made by a specialist. In America, for instance, it is frequently stated that *T. solium* is more common than *T. saginata*, but this view has been shown to be erroneous (Stiles, 1895, p. 281). According to the official medical statistics of the late civil war, 566 cases of tapeworms were noticed in 5,548,854 patients from July 1, 1862, to June 30, 1866, or 1 : 9,803, but no indication as to the species found is given. In some countries statistics seem to show that tapeworm disease has been on the increase. Thus Bérenger-Féraud (1892) records the following statistics for the maritime hospitals of France:

Year.	Cases.	Patients.	Cases per 1,000 patients.
1861–65	33	130,927	0.20
1866–70	95	152,822	0.62
1871–75	422	137,361	3.06
1876–80	1,108	130,898	8.45
1881–85	1,565	155,646	10.05
1886–90	2,253	152,352	14.80

Bérenger-Féraud looks upon 1860 as the date of introduction of *T. saginata* into France, but Blanchard has shown that this is not the case, although he admits that it has increased in frequency from year to year.

Krabbe has published the following valuable statistics regarding tapeworms of man in Denmark:

Year.	T. saginata.	T. solium.	Dipylidium caninum.	Bothriocephalus latus.
Before 1869	37	53	1	9
1869–80	67	19	4	11
1880–87	87	5	4	5
1887–95	89	....	6	30

It seems quite well established that there has been an increase in the frequency of *T. saginata* in man in some districts, but since Hertwig's important observation in 1889-90 regarding the seats of predilection of the larval stage the destruction of so many more larvae must necessarily have resulted in decreasing the frequency of this species in man. There can be no question that since the trichina scare in 1860 and the following years, which led to an inspection of pork in some countries, and to greater care in cooking it in others, *T. solium* has decreased in frequency.

The following key will aid in the determination of the tapeworms of man:

KEY TO THE ADULT TAPEWORMS OF MAN.

· [For forms recorded in this country follow Roman type.]

(1) Head with two elongate grooves or slit-like suckers; rostellum absent; uterus with special pore; genital pores generally dorsal or ventral.

Bothriocephalidae, 2.

Head with four cup-shaped suckers; rostellum present but not always evident; uterus without special pore; genital pores generally marginal... *Taeniidae*, 4.

BOTHRIOCEPHALIDAE (Subfamily *Bothriocephalinae*).

(2) Body with external segmentation; head with two elongate or groove-like suckers: Genital organs single in each segment; cirrus, vulva, and uterus open ventro-median .... *Bothriocephalus*, 3.
Genital organs double in each segment; cirrus, vulva, and uterus open ventrally; worm very large, attains about 10 meters in length by 2 cm. in breadth; life history unknown. Found in Japan .... Krabbea grandis.

BOTHRIOCEPHALUS.

(3) Very large, attains 10 meters or more in length, reddish gray in color; very rare in this country; obtained from eating fish: Common pike (*Lucius lucius*), ling (*Lota lota*), perch (*Perca fluviatilis*); several members of the salmon family (*Salmo umbla, S. trutta, S. lacustris, Thymallus vulgaris, Coregonus lavaretus, C. albula, Onchorrhynchus Perryi,* and perhaps *Salmo salar*) .... *B. latus.*
(*B. latus includes* B. cristatus *Daraine, 1874.*)
Length a little less than four feet; found in Greenland .... B. cordatus.
(*A larval Bothriocephalus* (B. Mansoni) *is found in subperitoneal connective tissue of man.*)

TAENIIDAE.

(4) Egg with thin outer shell and thick brown inner shell (embryophore); uterus median and longitudinal with lateral branches; head generally armed; larval stage a Cysticercus, Coenurus or an Echinococcus generally in herbivora; adults in carnivorous or omnivorous animals .... *Taeniinae*, 5.
Egg with thin transparent shells, and frequently in egg capsules; in some cases scattered through the segment; head nearly always armed with hooklets on rostellum; larval stage a cysticercoid; adults in birds and mammals.
Dipylidiinae, 7.

TAENIINAE.

(5) Head with armed rostellum .... 6.
Head unarmed, rostellum absent; strobila attains 3 to 10 meters in length; ovary of pore side undivided; uterus with 17 to 30 branches on each side; the most common tapeworm of man in this country; larva in cattle.
Taenia saginata, p. 71.
(6) Rostellum with two rows of hooks, 24 to 32 in number; strobila attains 4 to 8. meters in length; ovary of pore side divided; uterus with 7 to 12 branches each side; comparatively rare in this country; larva in swine.
T. solium, p. 89.
Rostellum (?)[1]; strobila attains 5 meters in length; terminal proglottids 27 to 35 mm. long by 3.5 to 5 mm. wide .... *T. confusa*.
(The larval stages of *T. solium* and *T. echinococcus* are also found in man.)

---

[1] Ward describes the rostellum as having 6 or 7 rows of very small hooks. Through the kindness of Professor Ward, I have recently examined the head and a number of segments of the original material. This examination leads me to look upon the head as a head of *Dipylidium caninum,* which has accidentally been placed in the wrong bottle, a possibility which had also occurred to Ward. Regarding the remarkable segments, I do not wish to commit myself until I have opportunity to study a complete specimen.

DIPYLIDIINAE.

(7) Suckers unarmed... 8.

Suckers armed (the suckers of the young specimens will undoubtedly be found to be armed, although the specimens thus far found in man were unarmed, the hooks probably having fallen), the hooks being arranged in circular rows on border; hooks on rostellum resemble a hammer in form, about 90 in number and arranged in a double row, or rostellum rudimentary and unarmed; strobila 25 to 30 cm. long. Very rare; not yet recorded for America. Larva probably in some invertebrate.
Davainea madagascariensis.

(8) Genital pores double; two submedian ovaries in each segment; several rows of hooks on rostellum; strobila attains 15 to 35 cm. in length; gravid segments elliptical. Adults found in dogs and cats; rare in man. Larva found in lice and fleas of dogs (*Trichodectes canis* and *Pulex serraticeps*).. *Dipylidium caninum.*

Genital pores single and unilateral (on left of segment); rostellum with 24 to 30 hooks, the dorsal root longer than prong or ventral root; three testicles normally present in each segment; eggs with three envelopes.... *Hymenolepis*, 9.

(9) Hooks (24 to 28 in number, 15 µ long) present on rostellum; body 10 to 15 mm. long; not uncommon in Italy; found also in other parts of Europe. Found in rodents (rats, etc.), as well as man; larva develops in the villi of the intestine .... *H. murina.*

(Including *Taenia nana*.)

Rostellum rudimentary and unarmed; 20 to 40 cm. or more long; adult generally parasitic in rodents (rats); larval stage develops in certain insects (*Asopia farinalis, Anisolabis annulipes, Akis spinosa, Scaurus striatus*)..... *H. diminuta.*

(Including *Taenia flavopunctata*.)

If a person is known to have a tapeworm, it is of great importance, both from an economic (agricultural) and a hygienic standpoint, to know whether one of these two tapeworms (*T. saginata*, the Beef-measle Tapeworm, and *T. solium*, the Pork-measle Tapeworm) is present, and if so, which one. This can be determined in several ways:

(1) Since the Pork-measle Tapeworm comes from pork, it will not be found in persons who abstain from eating that meat, as is usually the case among Hebrews. So that if one of these two parasites were found in such persons it could be only the Unarmed Tapeworm.

(2) On the other hand, those who eat pork but no beef would not be infested with the Unarmed Tapeworm, but we would expect to find in them the Pork-measle (armed) Tapeworm.

These two modes of diagnosis do not hold in all cases, for, as already stated, man is subject to ten different species of tapeworms. Most of the remaining eight species are, however, so totally different from the two under consideration that a positive diagnosis can be made by comparing the worms with figs. 73 and 81, and with the key given above.

(3) When segments of the parasite break off and wander out of their own accord or with stools of the person affected, it should be noticed whether several segments are joined together or whether every portion consists of a single segment. In the former case, the parasite is generally the armed parasite, in the latter case generally the unarmed parasite.

(4) Take a segment of the parasite found in the stool or bed, press it between two pieces of glass and hold it up to the light. Comparing

it with figs. 73 aud 81, notice whether the uterus has 17 to 30 branches on each side of the main trunk (Unarmed Tapeworm) or from 7 to 10 branches each side (Armed Tapeworm).

(5) In case the head is found, notice whether hooks are absent (unarmed) or whether two rows of hooks are present (armed).

The necessity of knowing which parasite is present is, first, that the Armed Tapeworm injures man, not only because it inhabits his intestine, but also because the larval stage may develop in the muscles, eye, and other portions of the body, and a man who has the Armed Tapeworm stands in constant danger of infecting himself with these larvae. The Unarmed Tapeworm, on the other hand, develops only in the intestine, its larval form being unable to develop in man.

In the second place, it is important to know which tapeworm is present in a person, especially in a farm hand, for if he has the unarmed form there is constant danger of his infecting the cattle by passing his excrements in fields where cattle feed; if he carries the Armed Tapeworm he will infect the hogs, should he void his excrements in a place to which swine have access.

Symptoms.—The symptoms exhibited by a patient troubled with tapeworms are both general and local: Itching at the extremities of the intestinal canal, and various dyspeptic symptoms; uncomfortable sensations in the abdomen,[1] uneasiness, fullness or emptiness, sensation of movement attributed to the movements of the parasite, colicky pains; disordered appetite, at times deficient, at other times craving; paleness and discoloration around the eyes; fetid breath; sometimes emaciation; dull headache; buzzing in the ears; twitching of the face; dizziness; often the uncomfortable feelings in the intestine are increased by fasting and relieved after a hearty meal; fainting, chorea, epileptic fits.

Diagnosis.—A positive diagnosis can be made by finding the segments of tapeworms in the stools, bed, or clothes of the patient, or by a microscopic examination of the faeces in search of eggs.

Treatment.—It is always advisable to consult a physician in regard to treatment, especially when the patient is much run down in health or naturally delicate in constitution, since "in weak persons, such as those having consumption, the treatment, if admissible at all, must be conducted with the greatest care, lest the patient's strength be exhausted" (Pepper). It is not always possible, however, for men on the ranches to obtain the services of a physician, so the following hints are given in regard to treatment and may be followed without danger by a strong and healthy person of ordinary intelligence:

Before taking any of the medicines suggested below, it is necessary to prepare for the treatment by removing all obstructions in the intestine to the free exit of the parasite. This can best be done by living

---

[1] The most constant symptom which I observed in an experimental infection of myself with *T. saginata* was the sensation which one experiences in the rapid descent of an elevator. This peculiar feeling frequently occurred, especially when walking.

for two or three days on a light diet of milk, coffee, soup, and bread; but vegetables should not be taken. On the evening before taking the medicine it is also advisable to give the patient a thorough injection of 1 to 2 quarts of warm water, to which 1 to 2 tablespoonfuls of pure glycerin may be added.

Early the next morning one of the following doses may be taken:

(1) Take 1 to 2 ounces of oil of turpentine + 1 ounce of castor oil, mixed with the white of an egg and some sugar. Take the whole dose at one time, and if a movement of the bowels does not follow within two or three hours, take another dose of castor oil (1 ounce) (Leidy, 1885).

Objection has been raised by some practitioners to the use of oil of turpentine on the ground that it causes an intense burning sensation in the intestine and produces headaches which may last several days.

(2) The most generally useful remedy is the oleo-resin of male fern,[1] which is in reality an ethereal extract of the drug. A half drachm or drachm of the remedy is given in the morning after two days' restriction of diet, and in the evening a brisk cathartic, such as castor oil, should be administered. Sometimes calomel is given in combination with the oleo-resin. The patient should remain abed after the administration of the remedy, to avoid syncope and other effects of large doses of the drug (Pepper, 1894). In overdose, this medicine is a distinct poison; six drachms have caused death.

(3) One or two ounces of pumpkin seeds ground and made into a paste with sugar. Follow in an hour with a dose of castor oil. This is one of the best, cheapest, and safest tapeworm remedies.

(4) Tanret's Pelletierine is very highly recommended but is rather expensive ($2.50 per dose) and often difficult to procure fresh in this country. In case this is taken the instructions which come with the bottle (one dose) must be strictly carried out.

Many other remedies could be suggested, but those given above are among the most simple and will suffice for this report.

Whatever anthelmintic is used, the medicine should be procured as fresh as possible. Many failures in treating for parasites are due to the fact that the remedy used has lost its anthelmintic property.

When the parasite is being passed the patient should evacuate into a vessel containing warm water, the object of this being to prevent the worm from breaking or attempting to retain its hold in the intestine in case it is still alive, as it will frequently do if it comes in contact with any cold object. The patient should likewise avoid pulling the worm while it is being expelled, for he is thus liable to break it.

When the movement is completed the stool should be examined thoroughly for the head, for if this has remained in the intestine it will give rise to new segments again, and in about three or four months the patient will discover that he is still infected. If the head is not found upon examination of the stool, it is best not to repeat the treatment until the segments have again appeared, for, as the head is quite small, it may have escaped notice, although present in the stool, and in that case the second treatment would be useless.

Prevention.—After what has been said, it is exceedingly easy to see

---

[1] Male fern and kamala capsules are put up ready for use. Directions come with each box.

the measures which should be adopted to prevent this disease: (1) Persons should not eat meat in which fresh cysticerci are present; (2) meat in which only a few cysts have been found, but have been cut out, should be thoroughly cooked or salted before eating, or, (3) such beef should lie in cold storage for three weeks at least; (4) cattle and hogs should not have access to human excrements, especially when it is known that persons in the neighborhood have tapeworms; (5) persons should not void their excrements on fields where live stock is feeding.

By following out these simple instructions it will not be a difficult task to totally eradicate the tapeworm disease caused by *T. saginata* and *T. solium* in man, and the corresponding disease of "measles" caused by the larvae of these worms in cattle and hogs. In fact, it has been noticed in several parts of Europe, where meat is inspected, that certain tapeworms are gradually becoming rarer, owing to the condemnation of meat containing the cysts.

20. Pork Measles (*Cysticercus cellulosae*) **of Man and Swine, and its adult stage, The Armed, or Pork Measle, Tapeworm** (*Taenia solium*) **of Man.**

[Figs. 75-83.]

Many authors state that the Pork-measle Tapeworm is the common tapeworm of man for the United States, but a careful study of the subject has shown this view to be erroneous.

LARVA (*Cysticercus cellulosae*).

For anatomical characters, compare figs. 75 and 76 with key, p. 21.

SYNONYMY.—*Finna* Werner, 1786; *Taenia hydatigera* Fischer, 1788; *T. cellulosae* Gmelin, 1790; *T. finna* Gmelin, 1790; *Vesicaria hygroma humana* Schrank, (-?-); *V. finna suilla* Schrank, (-?-); *V. lobata suilla* Fabricius, (-?-); *Hydatis finna* (Werner) Blumenbach, (-?-); *H. humana* Blumenbach, (-?-); *Taenia muscularis* Jördens, 1802; *T. hydatigena anomala* Steinbuch, (1802); *Cysticercus finna* (Gmelin) Zeder, 1803; *C. cellulosae* (Gmelin) Rudolphi, 1808; *C. finnus* (Gmelin) Laennec, 1812; *C. solium* Koeberlé, 1861; *C. suis* Cobbold, 1869; *Neotaenia* Sodero, (1886); *C. cellulosus* of several authors; "*Cysticerkus*" *cellulosae* of Schneidemühl, 1896.

ANOMALIES.—The names proposed by various authors for these supposed distinct species found in man, especially in the cranial cavity, are more or less descriptive. *Hydatis piriformis* Fischer, 1789 (=*Taenia pyriformis* (Fischer) Treutler, 1793=*Cysticercus pyriformis* (Treutler) Zeder, 1803 = *C. Fischerianus* Laennec, 1812); *Taenia albopunctata* Treutler, 1793 (= *C. albopunctatus* (Treutler) Zeder, 1803 = "*T. albopunctata hominis* Treutler" of Cobbold, 1864); *Cysticercus dicystus* Laennec, 1812; *C. acanthotrias* Weinland, 1858; *C. turbinatus* Koeberlé, 1861; *C. melanocephalus* Koeberlé, 1861; *C. racemosus* Heller, 1875 (=*C. bothryoides* Heller, 1875 [nec Reinitz, 1885] = *C. multilocularis* Küchenmeister, (-?-); *Trachelocampylus* Davaine, 1880 (for *Trachélocampules* Fredault, 1847).

HOSTS.—Man, swine, wild boar, and other animals. (See pp. 137-143.)

ADULT (*Taenia solium* Linnaeus, 1758).

For anatomical characters, compare figs. 77-81 with key, p. 84.

SYNONYMY.—*Taenia solium* Linnaeus, 1758 (after elimination of *T. saginata* and *T. marginata*); *T. cucurbitina* Pallas, 1766 (= *T. solium* Linnaeus, renamed); *T. cucurbitina* Art [=var.] *pellucida* Goeze, 1782; *T. cucurbitina, plana, pellucida* Goeze,

1782; *T. solitaria* Leske, (1785), pro parte; *Halysis solium* (Linnaeus) Zeder, 1803; *Taenia humana armata* Rudolphi, 1810, pro parte (= Brera's, 1802, *Tenia armata umana*); *T. solium* Linnaeus of Küchenmeister, 1852; *T. hamoloculata* Küchenmeister, 1855 (possibly earlier); *T. turbinata* Koeberlé, 1861; *T. (Cystotaenia) solium* of Leuckart, 1863; *T. tenella* Cobbold, 1874 [nec Pallas, 1781]; (?) *T. solium* var. *minor* Guzzardi Asmendo, 1876; *T. officinalis* Bos, 1894.

ANOMALIES.—(?) "*Taenia vulgaris*" Werner, 1782 [nec Linnaeus, 1758] = *T. dentata* Batsch, 1786; (?) *T. fenestrata* Chiaje, 1833; *T. (Cysticercus) acanthotrias* of Leuckart, 1863; (?) *T. fenestrata* Colin, 1885; *T. solium fenestrata* Colin, 1876; (?) *T. fusa*, *T. continua*, *T. solium fusa* seu *continua* Colin, 1876; (?) *T. scalariforme* Notta, 1885 (= *T. fenestrata* Colin, renamed) = *T. solium scalariforme* Notta, 1885.

PRE-LINNAEAN NAME.—(?) *Taenia degener* Spigelius, 1618.

BIBLIOGRAPHY.—For bibliography, see Huber, 1892. For technical discussion, see Leuckart (1880, pp. 617–713); R. Blanchard (1886, pp. 382–418).

HOST.—Man. It is an error for the Minnesota State board and the North Carolina Station to record it in dogs.

FIG. 75.—A piece of pork heavily infested with pork measles (*Cysticercus cellulosae*), natural size (original).

Life history.—The life cycle of the Pork Bladder Worm is exactly the same as that of the Beef Bladder Worm (see p. 72), except that the hog is the intermediate host. The following observations regarding the larval parasite at different ages have been made by various authors:

Nine days after infection.—An oval vesicle 33 μ long by 24 μ broad; connective tissue cyst absent. (Mosler.)

Twenty days after infection.—Parasite consists of a delicate, transparent bladder worm about as large as the head of a pin. The anlage (primordium) of the head is represented by a small, indistinct point; surrounding cyst absent. (Gerlach.)

Twenty-one days after infection.—Spherical, 0.8 mm. in diameter; slightly attenuate toward the point, showing the anlage (primordium) of the head. (Leuckart.)

Thirty-two days after infection.—Ellipsoid, 1 mm. to 6 mm. long by 0.7 mm. to 2.5 mm. broad. The largest specimens show the excretory system; the anlage (primordium) of the head equatorial; connective tissue cyst very thin.

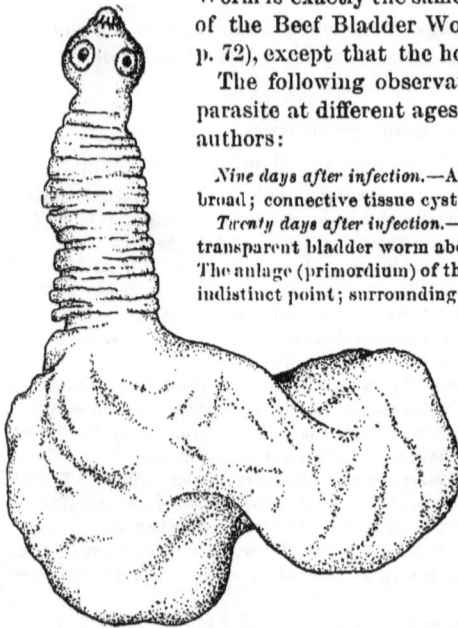

FIG. 76.— An isolated Pork-measle Bladder Worm (*Cysticercus cellulosae*), with extended head, greatly enlarged (original).

Forty days after infection.— Surrounding cyst still very delicate; about as large as a mustard seed, or somewhat larger; suckers and hooks visible, but not complete. (Gerlach.)

Sixty days after infection.—Size of a pea or slightly larger. When freed from the connective tissue, cyst somewhat renal in form; head as a small, white knob, but without neck; hooks and suckers fully developed. (Gerlach.)

One hundred and ten days after infection.—Neck developed; transverse lines slightly visible; head is invaginated in the bladder. (Gerlach.)

The determination of the age of the parasite is of importance in case that there is a guaranty of freedom from infection. According to Rasmussen the periods of guaranty are: In Prussia, Bavaria, and Austria, 8 days; in certain other parts of Europe, 9, 15, and 21 days; in Baden and Würtemburg, 28 days, and in Saxony, 30 days.

It is generally estimated that three to four months are required for the parasite to complete its development, but as the hooks and suckers are formed after two and a half months it is not impossible that a parasite ten to eleven weeks old would develop into the adult tapeworm if eaten by man; although, according to Gerlach, pork measles less than two months old are not dangerous. The longevity of the bladder worm varies with circumstances, but the factors here concerned are not understood. According to Railliet, cases have been observed in man where the bladder worm has caused severe cerebral troubles for twelve to fifteen years, and it has been observed in the eye for twenty years. The worms may undergo calcareous degeneration very early, but as a rule this does not take place until the cyst is quite old.

The degeneration begins with the capsule and ends with the scolex, and, according to Morot, may be divided into four stages, as follows:

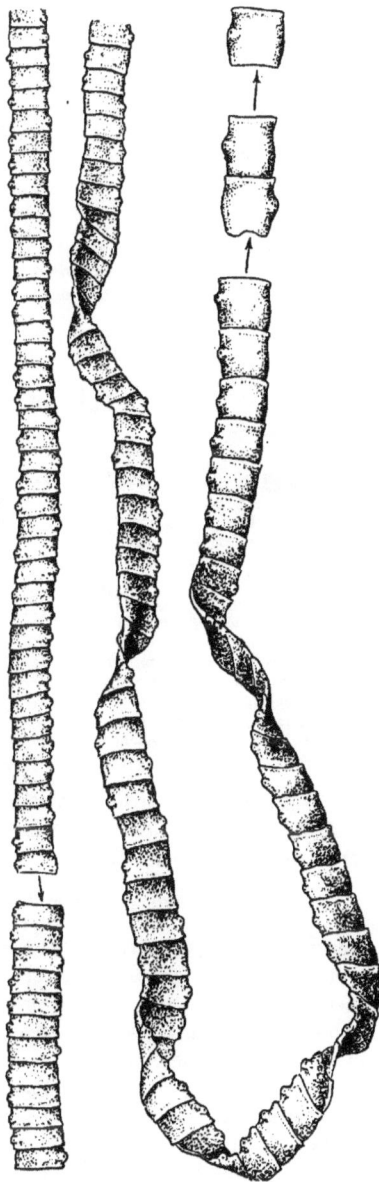

Fig. 77.—Several portions of an adult Pork-measle Tape worm (*Taenia solium*), natural size (original).

First stage.—The capsule shows cheesy, opaque spots, but the fluid is clear, and the scolex is intact.

Second stage.—Both the connective tissue capsule and the bladder cyst become cheesy; hooks are present, but the suckers more or less degenerated.

Third stage.—Hooks are present, but not in definite order or number.

Fourth stage.—No traces of the hooks can be found in the parasite, which is reduced to a cheesy mass.

Ostertag states that degeneration may take place before the hooks

FIG. 78.—Large (*a*) and small (*b*) hooks of Pork-measle Tapeworm (*Taenia solium*). ×280. (After Leuckart, 1880, p. 661, fig. 293.)

have formed. Ostertag has also shown that the hooks become loose, upon expression of the scolex, only in dead bladder worms.

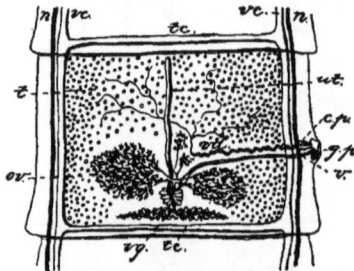

FIG. 79.—Mature sexual segments of Pork-measle Tapeworm (*Taenia solium*), showing the divided ovary on the pore side: *cp*, cirrus pouch; *gp*, genital pore; *n*, nerve; *ov*, ovary; *t*, testicles; *tc*, transverse canal; *ut*, uterus; *v*, vagina; *vc*, ventral canal; *vd*, vas deferens; *vg*, vitellogene gland. ×10. (After Leuckart, 1880, p. 665, fig. 294.)

PORK MEASLES.

The disease in hogs.—The symptoms in hogs are very indefinite, but a diagnosis may sometimes be made by examining the visible mucous membranes of the mouth, particularly under the tongue. See also the same subject for cattle, p. 77.

A heavy infection of measles is more common in hogs than in cattle—a fact easily understood when we recall the feeding habits of the two animals, the comparative size of their bodies, and of their stomachs. From 1884 to 1887, of 5,610 measly hogs found at the Berlin (Prussia) abattoir, 2,167 were heavily infested, 1,641 had medium infections, and 1,802 were slightly infested.

Treatment.—See page 77.

ABATTOIR INSPECTION.

See discussion, page 77. As the Armed Tapeworm is more dangerous to man than the unarmed form, the abattoir inspection for *Cysticercus cellulosae* is more important from a hygienic standpoint than the inspection for *C. bovis*.

Position of the parasites in hogs.—The Pork-measle Bladder Worm is found in the muscles, especially in the abdominal muscles, the muscular portion of the diaphragm, the psoas, tongue, heart, the muscles of mastication, intercostals,

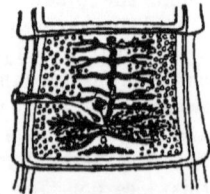

FIG. 80.—Segment of Pork-measle Tapeworm (*Taenia solium*) in which the uterus is about half developed. ×2. (After Leuckart, 1880, p. 666, fig. 295.)

muscles of the neck, the adductor of the hind legs, and the pectorals. These parts are shown by fig. 83. Particular stress should be laid upon an examination of the tongue and muscles of mastication, the muscles of the shoulder, neck, and diaphragm.

The parasite of hogs which is most liable to be mistaken for *C. cellulosae* is the large bladder worm with a long neck (*C. tenuicollis*, p. 96). The latter form, which is not transmissible to man, occurs under the serous membranes of the body cavities, but is not found in the muscles; it is much larger, and is provided with more hooks (28 to 40) than *C. cellulosae* (22 to 28).

Frequency of Cysticercus cellulosae in hogs.—Satisfactory statistics regarding the presence of this parasite in American hogs are lacking; we know, however, the American hogs are comparatively free from this worm. The statistics for Prussia are quite complete, the proportion of infested hogs being as follows:

1876 to 1882.—One hog infested in every 305 hogs examined. (Johne, after Ostertag.)
1886 to 1893.—One hog infested in every 637.5 hogs examined.[1]

The proportion of measly hogs appears to vary in different localities. Thus at the Berlin abattoir the average for seven years was 1 infested hog to every 173 hogs examined (Ostertag). In south Germany the parasite is said to be rare. It is much more common in the eastern Prussian provinces than in the western, as shown by Ostertag in the following statistics for 1892:

Regierungsbezirk.	Propor-tion.	Regierungsbezirk.	Propor-tion.
Marienwerder	1:28	Arnsberg	1:865
Oppeln	1:80	Coblenz	1:975
Königsberg	1:108	Düsseldorf	1:1070
Stralsund	} 1:187	Münster	} 1:1930
Posen		Wiesbaden	
Danzig		Entire Prussia	1:1290
Frankfurt	} 1:250	Eastern provinces	1:604
Bromberg			

The hogs imported into Germany from Russian Poland, Galicia, Bohemia, and Siberia were infested in much higher proportion than the German hogs; in some of the importations the proportion ran as high as 50 per cent (Ostertag).

[1] The totals of the following table do not agree with the totals published in Germany, but are made upon the details given for the various years. Several errors in addition were noticed in the German statistics.

Year.	Hogs inspected.	Hogs in-fested with *C. cellulosae.*	Authority.
1886	4,834,898½	10,126	} Veröff. d. kais. Gesundheits-amtes, 1891, p. 244.
1887	5,486,416½	11,068	
1888	6,051,249½	10,031	
1889	5,500,678½	8,373	
1890	5,590,512	5,492	} Veröff. d. kais. Gesundheits-amtes, 1894, p. 208.
1891	6,550,182	7,689	
1892	6,134,550	9,364	
1893	6,251,776	10,640	Veröff. d. kais. Gesundheits-amtes, 1895, p. 347.
Total	46,400,272	72,783	

Influence of the age of the host.—According to Gerlach hogs over six months old will not become infested with this parasite, but this is not admitted by Fischöder.

Disposition of measly pork.—See disposition of measly beef, p. 81. Since *Taenia solium* is more dangerous than *T. saginata*, the regulations concerning the disposition of measly pork should be even more rigid than those concerning measly beef. Pork of this character can of course be sold under declaration, but even this is not advisable unless the meat is first rendered non-infectious. Cases of so-called "single infection"[1] should be treated the same as cases of moderate infection. In very heavy infections (up to 20,000 bladder worms may occur in a single carcass) the

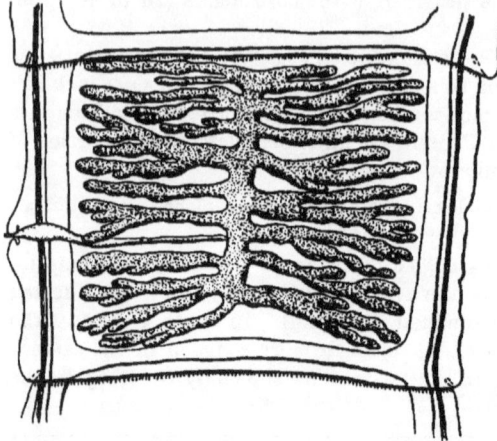

FIG. 81.—Gravid segment of Pork-measle Tapeworm (*Taenia solium*). showing the lateral branches of the uterus, enlarged (original).

pork is watery and pale; it decomposes easily, and has a disagreeably sweet taste. Such cases should, of course, be condemned to the tank.

Cooking.—According to Perroncito, *C. cellulosae* dies at 45° to 50° C. (= 113° to 122° F.). (See also p. 81.) It dies in 1 minute at 50° C.

Storage.—Living specimens of *C. cellulosae* have been found in pork twenty-nine days after slaughtering (Railliet). After fourteen to nineteen days of cold storage at −10° to −15° C. the parasites are said to be dead; the protoplasm has become viscid, bluish opaque, and the hooks have fallen. More observations are needed upon this subject.

Effects of electricity.—Glage has experimented some with electricity in order to kill the parasite of pork measles, but

FIG. 82.—Eggs of Pork-measle Tapeworm (*Taenia solium*): a, with primitive vitelline membrane; b, without primitive vitelline membrane, but with striated embryophore. ×450. (After Leuckart, 1880, p. 667, fig. 297.)

further study in this line is desirable before the method is adopted.

THE ADULT AND LARVAL TAPEWORM IN MAN.

See discussion, page 83. It should always be recalled that the Armed Tapeworm is more dangerous than the Unarmed Tapeworm, since, as already stated, the larva as well as the adult may develop in man.

---

[1] During the year 1889–90 Berlin, Prussia, found 373 cases of alleged single infection; of these, 56 cases were afterward shown to contain more than one parasite.

Treatment.—See page 87.

Cysticercus cellulosae *in man.*—This infection may take place in different ways; a patient may either soil his hands with the microscopic eggs during defaecation and afterward swallow the eggs; or, through a reverse peristaltic movement of the intestine, gravid segments may be carried into the stomach, where the shells will be destroyed, thus freeing the embryos. An infection through a contaminated water supply may also take place. (See Life history, p. 90.) In man the bladder worm may develop in the muscles, the eye, and the brain.

The following statistics upon the distribution of the worm in various parts of the body have been compiled from different sources:

Locality.	Parasite found in—						Total number of cases.	Authority.
	Brain.	Muscles.	Heart.	Lungs.	Under skin.	Liver.		
Dresden	21 {	11	3				22	}Müller.
Erlangen		1					14	
Berlin	72	13	6	3	3	2	87	Dressel.
Erlangen	13	6			2		21	Hang (1874-1885).
(?)........	6	1					6	Gribbohm.
(?)........	5						5	Sievers.
Total	117	32	9	3	5	2	155	

The following statistics refer to the presence of this parasite in the eye:

Locality.	Total number of patients or bodies.	C. cellulosae in the eye.	Authority.
Berlin .... {	80,000	80	von Gräfe.
	60,000	70	Hirschberg (1869-1885).
Berlin .... {	30,000	1	Hirschberg (1886-1889).
	43,000	2	Hirschberg (1890-1894).
France ....|	60,000	1	De Wecker }Parasite in crystallin lens.
Austria................	30,000	7	Mauthner

According to Virchow, the proportion of cysticercus in the human cadavers dissected in Berlin has been reduced from 1:31 (before the introduction of meat inspection) to 1:280 (since the introduction of meat inspection).

The following statistics, collected from various sources by Blanchard, refer to post-mortem examinations:

Locality.	Number of post-mortems.	Number infected.	Proportion infected.
Switzerland:			*Per 1,000.*
Zürich	2,500	1	2.5
Basel..	1,100	0	.0
Basel..	1,914	6	1.13
Germany:			
Kiel...			6
Erlangen..			6.7
Dresden...			11.3
Berlin..			16.4
Berlin..			12.5

Prevention.—See page 88.

21. The Thin, or Long, Necked Bladder Worm (*Cysticercus tenuicollis*) **of Cattle, Sheep, and Swine, and its adult stage, The Marginate Tapeworm** (*Taenia marginata*) **of Dogs and Wolves.**

[Figs. 84–87A, 88, 89B, 90–93.]

Still another bladder worm, which is by no means uncommon in the animals of this country, occurs in the body cavity of cattle, sheep, swine, and other animals, attached to the diaphragm, omentum, liver, or other organs. When eaten by dogs or wolves, it develops into the Marginate Tapeworm, which was formerly confused with *T. solium* of man, and gave rise to the erroneous idea that the Pork-measle Tapeworm occurs in dogs as well as in man.

LARVAL STAGE (*Cysticercus tenuicollis*).

For anatomical characters, compare fig. 84 with key, p. 21.

SYNONYMY.—*Taenia hydatoidea* Pallas, 1760; *T. hydatigena* Pallas, 1766, pro parte; *Hydra hydatula* Linnaeus, (1766); *Vermis vesicularis eremita* Bloch, 1782; *Hydatigena orbicularis* Goeze, 1782; *H. globosa* Batsch, 1786; *H. oblonga* Batsch, 1786; *Vesicaria orbicularis* Schrank, 1788; *Taenia simiae* Gmelin, 1790; *T. ferarum* Gmelin, 1790; *T. caprina* Gmelin, 1790; *T. orilla* Gmelin, 1790; *T. verrecina* Gmelin, 1790; *T. bovina* Gmelin, 1790; *T. apri* Gmelin, 1790; *T. globosa* (Batsch) Gmelin, 1790; *Hydatula solitaria* Viborg, (1795); *Cysticercus claratus* Zeder, 1803; *C. simiae* (Gmelin) Zeder, 1803; *C. caprinus* (Gmelin) Zeder, 1803; *C. tenuicollis* Rudolphi, 1810; *C. visceralis simiae* Rudolphi, 1810 (*T. simiae* Gmelin, renamed); *C. lineatus* Laennec, 1812; *C. ovis* Cobbold, 1865; *Monostomum hepaticum suis* Willach, 1893; "*Cysticerkus*" *tenuicollis* of Schneidemühl, 1896.

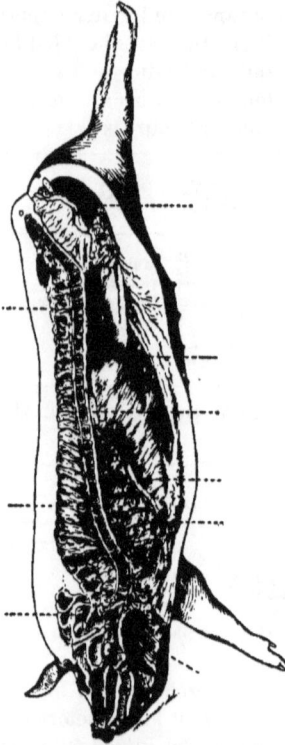

FIG. 83.—Half of hog, showing the portions most likely to become infested with pork measles. (After Ostertag, 1895, p. 387, fig. 79.) See p. 92.

PRE-LINNAEAN NAMES.—*Hydatides* Bartholini, 1673; *Vermes vesiculares* Hartmann (1685), quoted by Pallas as *Hydatis animata*; *Lumbricus hydropicus*, Tyson, 1691.

HOSTS.—Cattle, sheep, swine, deer, and other animals. (See pp. 137–143.)

ADULT STAGE (*Taenia marginata* Batsch, 1782).

For anatomical characters, compare figs. 85–89 with key, p. 101.

SYNONYMY.—See also pp. 89–90. *Taenia solium* Linnaeus, 1758, pro parte; *T. cateniformis* Goeze, 1782, pro parte; *T. marginata* Batsch, 1782; *T. lupina* Schrank, (1788); *T. cateniformis β. lupi* Gmelin, 1790; *Halysis marginata* (Batsch) Zeder, 1803; also "*T. solium*" of dogs, of several medical authors.

BIBLIOGRAPHY.—For technical discussion, see Deffke, 1891.

HOSTS.—Dog and wolf. (See pp. 137–143.)

Life history.—In tracing the life history it is best to begin with the egg, produced by the adult tapeworm in the intestine of dogs. These eggs, containing a six-hooked embryo, escape from the dog with the excrements and are scattered on the ground, either singly or confined in the escaping segments of the tapeworm. Once upon the ground they are easily washed along by rain into the drinking water, ponds, or brooks, or scattered on the grass. Upon being swallowed with fodder or water, they arrive in the stomach of the intermediate host (cattle, sheep, etc.), where the eggshells are destroyed and the embryos set free. The embryos then traverse the intestinal wall, and according to most authors arrive either actively, by crawling, or passively, by being carried along by the blood, in the liver or lungs, where they undergo certain transformations in structure. While still in the finer branches of the blood vessels of the liver, which they transform into small irregularly shaped tubes about 12 to 15 mm. long and 1 to 1.5 mm. broad, the embryos lose their six hooks, and develop into small round kernels, which are generally situated at one end of the tubes. The embryo can first be seen about four days after infection. The "scars" (figs. 91 and 92) described in the liver of animals infested with *Cysticercus tenuicollis* are nothing more nor less than these tubes, or altered blood vessels, caused by the growth and wandering of the parasites. In a shoat which Leuckart infected with eggs, and which he killed twenty-three days after the infection, he found two young cysticerci in the liver 6 to 8.5 mm. long and 3.5 to 5 mm. broad. In the smaller parasite no head was visible; in the larger, one end was slightly differentiated and evidently represented the anlage (primordium) of the

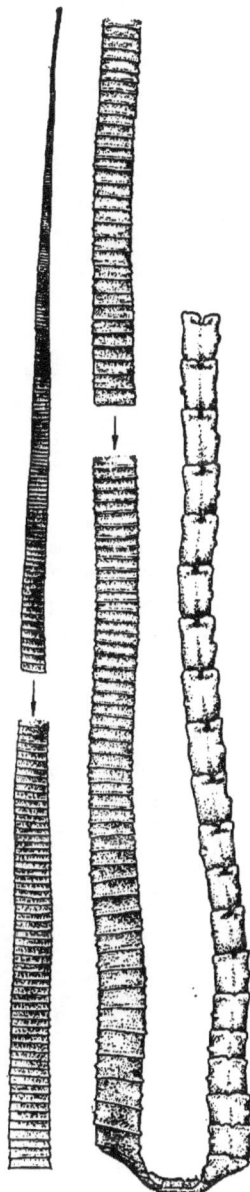

FIG. 84.—The Thin-necked Bladder Worm (*Cysticercus tenuicollis*), with head extruded from body, from cavity of a steer, natural size (original).

FIG. 85.—The Marginate Tapeworm (*Taenia marginata*), natural size (original).

scolex—that is, the head and neck in course of development. The portion which was destined to give rise to the head and neck was a small projection extending into the cavity of the hydatid. At about this stage, or a few days later, the parasites leave the liver, fall into the body cavity, and become encysted again in the organs mentioned above.

A month after infecting another shoat, Leuckart found cysticerci in the body cavity, with partially developed suckers and hooks. Six weeks after the infection of another shoat, he found cysticerci 15 mm. long encysted in the omentum, and with fully developed scolex. Three months after infecting a lamb, he found cysticerci twice as large. Experiments have also been made by other authors (Baillet, Küchenmeister, Railliet, etc.), most of them agreeing with Leuckart's experiment.

FIG. 86.—Head of the Marginate Tapeworm (*Taenia marginata*). × 17. (Original.)

Curtice, however, takes a somewhat different view, that is, he considers the liver as a place of destruction for the young parasites, rather than a normal place for their development; he also claims that the embryos which may even travel the entire length of the intestine of the intermediate host, traverse the intestine and arrive directly in the position where they complete their larval development without first passing through the liver.

After developing into the full-grown bladder worm, the parasites remain unchanged until they are devoured by a dog or wolf, or until,

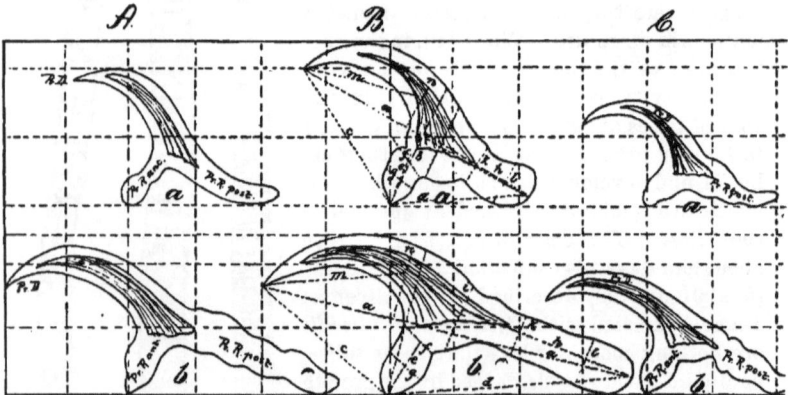

FIG. 87.—Small and large hooks of (A) (*Taenia marginata*), (B) *T. serrata*, and (C) *T. cornurus: a*, small hooks; *b*, large hooks. × 480. (After Deffke, 1891, Pl. II, fig. 9.)

after an undetermined length of time, they become disintegrated and more or less calcified.

If the hydatid is devoured by a dog or wolf, either when the latter prey upon the secondary host or when the dog obtains the cyst at a slaughterhouse, the bladder portion is destroyed, the scolex alone remaining intact in the digestive fluids. The head holds fast to the intestinal wall with its suckers and hooks; by strobilation (transverse division) it

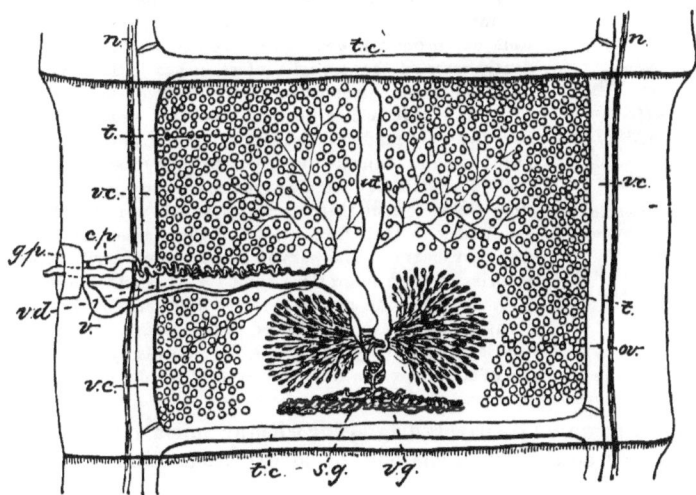

FIG. 88.—Sexually mature segment of the Marginate Tapeworm (*Taenia marginata*): *cp*, cirrus pouch; *gp*, genital pore; *n*, nervo; *ov*, ovary; *sg*, shell gland; *t*, testicles; *tc*, transverse canal; *ut*, uterus; *v*, vagina; *vc*, ventral canal; *vd*, vas deferens; *vg*, vitellogeno gland. Enlarged. (After Deffke, 1891, Pl. I, fig, 1.)

gives rise to the segments, which, as we have already seen, together with the head, go to make up the adult tapeworm. Reproductive organs of both sexes develop in the separate segments, and eggs are

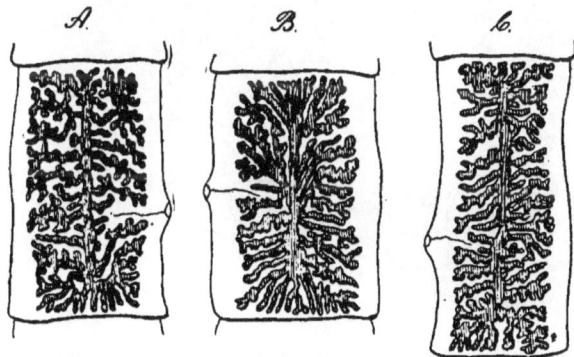

FIG. 89.—Gravid segments, showing the lateral branches of the uteri of (A) *Taenia serrata*, × 4; (B) *T. marginata*, × 6; (C) *T. coenurus*, × 10-15. (After Leuckart, 1880, p. 720, fig, 308.)

produced within which are developed the six-hooked embryos, the point from which we started out.

The disease in cattle, sheep, and hogs.—As a rule, this bladder worm is a comparatively harmless parasite, a light infection having little or no effect upon the host.   A heavy infection may, however, prove fatal to young animals.

So far as I know, only one case is on record where this parasite has proven fatal to cattle—probably from the fact that no severe infections have yet been found; and from our present knowledge of the subject, it can be confidently asserted that a slight infection has little or no effect upon this host. In experiments which have been made upon sheep and pigs, it has been noticed that heavy infections have not only produced decided symptoms but have proved fatal to the animals named.   Several cases (Leuckart, Zschokke, Railliet) are also reported where pigs have died from the effects of these parasites which were accidentally acquired with their food. In all of these cases the infection was very heavy. The parasites had caused peritonitis and pleurisy by their migrations from the liver and lungs to the body cavities. In a case recently described by Railliet, a shoat of two months succumbed to the disease. Baillet made numerous experiments on lambs and on young goats, the animals dying in ten to fifteen days (a primordium of the scolex was noticed on cysts fifteen days old). In one of Railliet's experiments a goat died in five days.

FIG. 90.—Egg of the Marginate Tapeworm (Taenia marginata) with six-hooked embryo, greatly magnified (original).

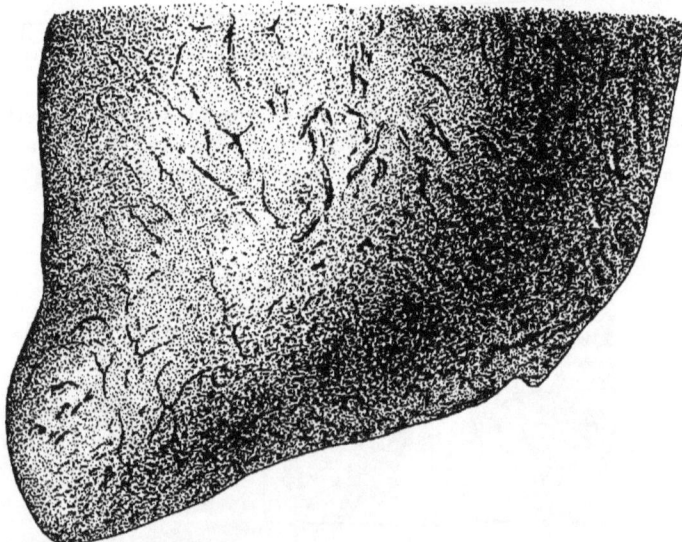

There is no way of positively diagnosing when an animal is infested with these larvae, as the symptoms noticed on experimental animals apply

FIG. 91.—Portion of the liver of a lamb which died nine days after feeding with eggs of the Marginate Tapeworm (Taenia marginata), with numerous "scars," due to young parasites. (After Curtice, 1890, Pl. X, fig. 1.)

equally well to other infections.   Diagnosis being uncertain or even impossible, it is useless to discuss treatment, except to remark that the parasite can not be reached with medicines; so that any treatment advised would be simply that advocated for pleurisy or peritonitis.

Prevention is a comparatively easy matter and lies in keeping dogs free from tapeworms.

ABATTOIR INSPECTION.

So far as the question of using beef, mutton, or pork from animals infested with *Cysticercus tenuicollis* as food for man is concerned, this parasite is of no importance whatever; for although several authors have attempted to infect themselves with tapeworms by swallowing this larva, all such experiments have been negative.

Differential diagnosis.—Infection of cattle, sheep, and hogs by *C. tenuicollis* may be mistaken for infection by *C. bovis* (p. 71), *C. cellulosae* (p. 89), *Echinococcus* (p. 113), and even for *tuberculosis*, but the differential diagnosis should not be difficult. For the differences between the Long-necked Bladder Worm and the other three larvae, see the discussions of those parasites. The condition of the corresponding lymphatic glands in tuberculosis of the host, as well as the hooks and calcareous

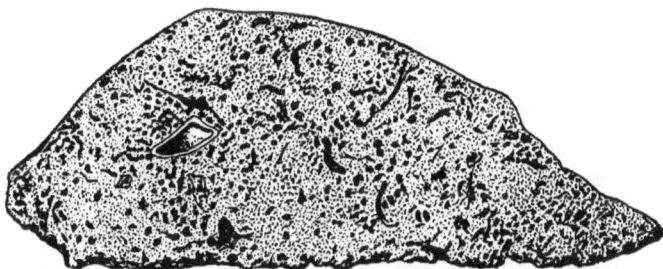

Fig. 92.—Cross section of the liver of a lamb which died nine days after feeding with eggs of the Marginate Tapeworm (*Taenia marginata*). (After Curtice, 1890, Pl. X, fig. 2.)

corpuscles of *C. tenuicollis*, allow a differentiation of degenerated and calcified specimens of this parasite from tuberculosis.

Frequency of C. tenuicollis.—In some countries the larval stage is common, especially in sheep. Olt found it in 26.4 per cent of the sheep (132 times in 500 sheep) examined at Stettin, Germany. It occurs in America, Europe, Africa, and probably elsewhere.

In connection with this parasite it is necessary to consider the adult tapeworms found in dogs.

THE ADULT TAPEWORMS OF DOGS.

The Marginate Tapeworm is unfortunately not the only tapeworm in the dog which proves harmful to our flocks. The following key will aid in determining the most common canine forms and show the source of infection:

KEY TO THE ADULT TAPEWORMS OF DOGS.

[For the forms transmissible to cattle, sheep, and swine follow Roman type.]

(1) Four suckers on the head .... Family *Taeniidae*, 2.
    Two suckers on the head; genital pores ventro-median .... *Bothriocephalus.*

(2) Head armed with hooks; genital pores marginal........................... 3.

Head not armed with hooks; genital pores ventro-median............. Mesocestoides.

(3) Head armed with a double row of hooks; genital pores on only one side of each segment.. 4.

Head very small, with about 60 hooks arranged in 4 rows; body 10 to 40 cm. long, with 80 to 120 quadrate to elliptical segments, the largest of which may measure 1.5 to 3 mm. broad by 8 to 10 mm. long; eggs in round capsules, about 250 in number, with 5 to 20 eggs in each capsule; eggs measure 43 to 50 μ; larval stage in fleas and lice, which may transmit the worm to dogs or to man........ Dipylidium caninum.

(4) Body small, 4 to 5 mm. long, with only 3 to 4 segments, the largest of which may measure 0.6 mm. broad by 2 mm. long; 28 to 50 hooks on the head; about 60 testicles present in a segment; embryophores 32 to 36 μ by 25 to 30 μ. The eggs are transmissible to man, oxen, sheep, pigs, horses, and other mammals, and develop into the larval stage (the Echinococcus hydatid), which is very dangerous. All dogs found infested with this worm should be killed and burned.. *Taenia echinococcus*, p. 113.

Body much larger and with many more segments............................ 5.

(5) Segments somewhat broader than long, or square, or longer than broad ...... 6.

Segments much broader than long, except the distal segments which suddenly elongate; head small, with 26 to 34 hooks; genital pore unusually large and prominent; embryophores 30 μ. Larval stage develops in the reindeer .... Taenia Krabbei.

(6) Ventral root of hooks simple.. 7.

Ventral root of smaller hooks bifid....................................... 8.

(7) Strobila 40 to 60 cm. long, rarely 1 m.; head pyriform, 0.8 mm. in diameter, with 22 to 32 hooks, the larger hooks 150 to 170 μ long; 220 to 250 segments present; distal 12 to 15 segments measure 8 to 12 mm. long by 3 to 4 mm. broad; 18 to 26 uterine branches (fig. 89 C) each side of median stem; about 200 testicles in each segment; embryophores spherical, 31 to 36 μ. Transmissible from dogs to lambs and calves, in which animals it causes "Gid".. *Taenia coenurus*, p. 103.

Strobila 1.5 to 5 m. long; head renal to square, 1 mm. broad, with 28 to 44 hooks, the larger hooks 188 to 220 μ long; 650 to 700 segments present; distal 50 to 70 segments measure 10 to 14 mm. long by 4 to 7 mm. broad; 5 to 6 or 8 uterine branches (fig. 89 B) each side of median stem; about 600 testicles in each segment; embryophores spherical, 31 to 36 μ. Transmissible from dogs to cattle, sheep, swine, etc.............................. *Taenia marginata*, p. 96.

(8) *Strobila 45 to 72 cm. long; head globular, 0.85 to 1.3 mm. in diameter, with 26 to 33 hooks, the larger of which measure 135 to 156 μ long; largest segments 8 to 16 mm. long with prominent posterior edge; embryophores ovoid, 33 to 41 μ by 26 to 31 μ. Transmissible to rabbits and hares............................. Taenia serialis.*

Strobila 60 cm. to 2 m. long; head 1 to 3 mm. in diameter, with 38 to 48 hooks, the largest measuring 225 to 250 μ long; about 400 segments present, of which the distal 30 to 40 measure 10 to 17 mm. long by 4 to 6 mm. broad; posterior edge of segments very prominent, giving strobila a serrate appearance; about 400 testicles in each segment; uterus with 8 to 10 lateral branches (fig. 89 A) each side of median stem; embryophores ovoid, 36 to 40 μ by 31 to 36 μ. Transmissible to rabbits and hares .... Taenia serrata.

Tapeworm disease in dogs.—It is the exception that the presence of tapeworms in dogs is diagnosed symptomatically, since in the majority of cases, especially in light infections, these parasites do not affect dogs to such an extent as to attract attention. Many a house dog or hunting dog harbors tapeworms without their presence ever being suspected. In some cases, however, the worms cause more or less serious pathological lesions in the intestine, which naturally bring about pronounced symptoms, although the direct cause of the trouble is not always apparent to the diagnostician.

The general symptoms exhibited are a change in appetite, disposition to vomit, general restlessness, occasionally cramps.

The smaller species of tapeworms, *Taenia echinococcus* and *Dipylidium caninum*, have been accredited with doing more harm than the larger forms (*T. marginata*, *T. serrata*, *T. coenurus*, and *T. serialis* and *Bothriocephalus*). A heavy infection of *T. echinococcus* may cause a severe, in some cases fatal, intestinal inflammation, with hemorrhage, the dog exhibiting epileptic symptoms or even symptoms which might be mistaken for hydrophobia—change of voice, tendency to bite, weakness, paralysis of lower jaw, etc. *Dipylidium caninum* occasionally bores tunnels in the mucosa of the intestine "through which the strobila is drawn, much like a train of cars." Schieferdecker found a peculiar hypertrophy of the intestinal villi in a dog infested with this parasite, the villi being four to five times the normal length; the glands of Lieberkühn were more or less atrophied. The same severe symptoms mentioned for infection with *T. echinococcus* have also been noticed in dogs infested with *D. caninum*.

An accumulation of tapeworms may result in a stoppage of the bowels, and cases are on record of perforation of the intestinal wall by *T. serrata*.

The nervous symptoms are more pronounced in high-strung dogs. Regarding frequency, it may be stated that tapeworms are more common in butchers' dogs and stray dogs having access to slaughterhouses than in other dogs. It is claimed by some authors that male dogs are more frequently infested than female dogs, and that tapeworms are more common in large dogs and in dogs from 1 to 3 years old than in small dogs and animals under 1 year of age.

Fig. 93.—Young cysticercus (*Cysticercus tenuicollis*) of the Marginate Tapeworm (*Taenia marginata*), natural size. (After Curtice, 1890, Pl. X. fig. 3a.)

The best method of diagnosis is to examine the faeces for segments or eggs. In some cases the attention of the diagnostician is attracted to expelled segments by the dog's licking around the anus or his "sliding" on the anus. A mild laxative will generally result in the expulsion of a few segments, and this method of confirming suspicions is occasionally used.

Even if segments are not found in the excreta, it is a good plan to treat the dogs for tapeworms, so as to remove all doubts as to their presence. Dogs which come in contact with herds should certainly be treated occasionally to prevent any possibility of infecting the stock animals.

It will be noticed that the larval stages of three of these tapeworms are injurious to stock animals, namely, *T. marginata*, *T. coenurus*, and *T. echinococcus*.

The larvae of the others are of comparatively little economic importauce, although it may be remarked that the larvae of *T. serrata* are sometimes fatal to rabbits, while the adult *D. caninum* sometimes occurs in children, who become infected with it by too intimate association with dogs. While playing with dogs they unconsciously get fleas upon themselves which they afterwards swallow. The fleas are digested and the larvae contained in their bodies, becoming free in the intestine, develop into tapeworms.

It is very difficult to distinguish between the adult forms of *T. coenurus* and *T. serialis*. The later is quite common in America.

If *T. echinococcus* is found to be present in a dog, the safest plan is to kill the dog and burn its carcass. The larval form of this parasite is so dangerous to man that it is not safe to have the dog around or to handle it, as is necessary in administering the treatment.

Taenia marginata develops in the dog, as stated elsewhere, in about ten to twelve weeks; *T. serrata* in about eight weeks; *T. coenurus* in two and a half to eight weeks.

The table following, giving the more common tapeworms in dogs, has been compiled from the various sources cited, and shows the comparative frequency of the various forms. There are as yet no extensive statistics for this country.

Number and percentage of dogs infested with tapeworms.

Locality.	Total number of dogs examined.	Taenia marginata.		Taenia serrata.		Taenia coenurus.		Taenia serialis.		Taenia echinococcus.		Dipylidium caninum.		Mesocestoides lineatus.		Bothriocephalus.		Authority.
		No.	Per cent.	No.	Per cent.	No.	Per cent.	No.	Per cent.	No.	Per cent.	No.	Per cent.	No.	Per cent.	No.	Per cent.	
Washington and elsewhere[1]	(?)	Present.		Present.		0	0	Present.		Very rare.		Common.		0	0	0	0	Stiles & Hassall (unpublished).
Washington, D. C.	50	1	2	6	12	0	0	0	0	0	0	22	44	0	0	0	0	Sommer, 1896.
Lincoln, Nebr.	20	1	5	9	45	0	0	1	5	0	0	13	65	0	0	0	0	Ward, 1897, p. 172.
Iceland, 1863	100	75	75			18	18			28	28	57	57	21	21	5	5	Krabbe, 1865, p. 21.
Copenhagen and vicinity, 1860–1863.	500	71	14.2	1	.2	5	1			2	.4	240	48			1	.2	Krabbe, 1865, p. 4.
Copenhagen.	(?)	(?)	17.3	(?)	15	(?)	2.16			(?)	1.08	(?)	47.03	(?)	(?)	(?)	(?)	Krabbe,[2] 1862 or 1852?
Leipzig, Saxony	(?)	(?)	27			(?)	1					(?)	25					Schöne, 1886.
Berlin, Prussia.	20	No record.		No record.		No record.		No record.		0	0	No record.		No record.		No record.		Nannyn, 1863, p. 415.
Berlin, Prussia, 1888.	200	14	7	10	5	1	.5	No record.		2	1	80	40			1	.5	Deffke, 1891.
Stettin.	12	No record.		No record.		No record.		No record.		3	25	No record.		No record.		No record.		Olt, after Ostertag, 1895, p. 424.
Zürich, Switzerland.	(?)			(?)	2.3	(?) 1	1.7			3.9		(?) 21		7.1					Zschokke.
Lyons and vicinity, France.	(?) 84	11	13	23	27.3	(?) 1	1.2			6	7.1	75	89.2	[3]6	7.1			Bertolus[4] & Chauveau, 1879, p. 309.
Melbourne, Australia, 1883	10	4	40							5	50	6	60			[5]1	5	J. D. Thomas, 1884, p. 192.
Adelaide, Australia.	20	[6]8	40							9	45	[6]18	90					J. D. Thomas, 1882, p. 436; 1884, p. 190.
Elsewhere in South Australia, 1882–81.	10	No record.								4	40	No record.						J. D. Thomas, 1884, p. 190.

[1] From time to time Hassall, Curtice, Stiles, and several volunteer assistants have examined dogs in this laboratory, but none of the examinations except those made by Sommer were conducted with a view to establishing statistics, and hence no exact records have been kept. I have personally seen adults of T. marginata, T. serrata, T. serialis, T. echinococcus, and D. caninum, and larvae of the four species of taenia collected in various parts of the United States.

[2] After Deffke, 1891, p. 259; Krabbe's article not in Washington.

[3] Recorded as " T. inermis " = T. pseudo-cucumerina = M. lineatus.

[4] Also observed one case of T. serialis (see B & C., 1879, p. 297), but apparently not in this lot of dogs.

[5] Thomas is not certain that these specific determinations were " invariably correct."

[6] This is recorded from the first 13 dogs examined (see Thomas, 1882, p. 456); no mention is made of the Bothriocephalus in the other dogs.

Treatment.[1]—The method of treatment is much the same as that followed in tapeworm disease of man; first prepare the patient by feeding him on a light diet of milk, soup, bread, etc., and then administer anthelmintics. It is important that the dog should be confined during the entire period of preparation and treatment.

In selecting a remedy, it is well to consider the following drugs. The

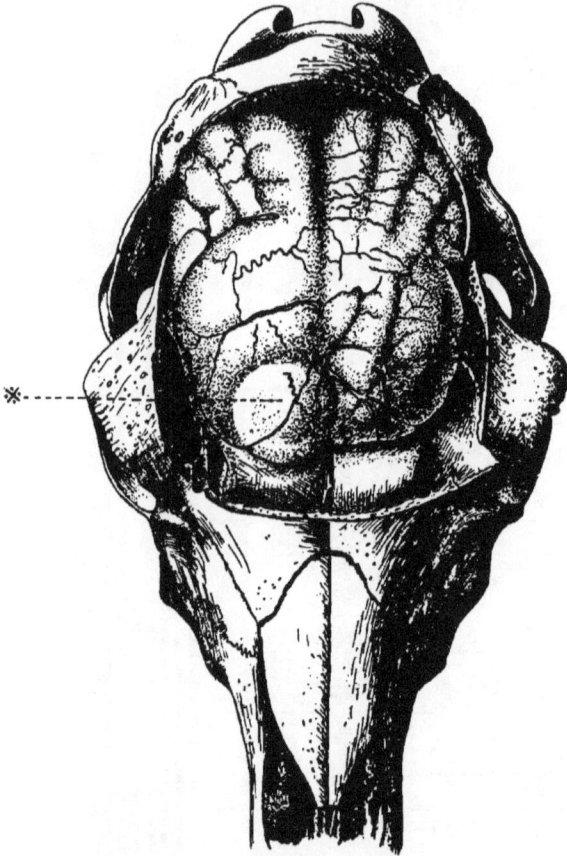

FIG. 94.—Skull of a sheep showing the brain infested with a Gid Bladder Worm" (*Coenurus cerebralis*), ½ natural size. (After Railliet, 1893, p. 256, fig. 150.) See p. 108.

doses (apothecaries' weight) here given and the remarks on the drugs are abstracted from French (1896).

The doses of pelletierine tannate are, for adults, 5 to 15 grains; puppies, ¼ to 5 grains. Pelletierine is undoubtedly the most efficient and innocuous taeniacide for the dog we possess, but is not much used on account of its expense. French

[1] In this connection consult French, 1896, and Curtice, 1890, pp. 77-78.

has frequently found it most useful when the stomach has refused to retain other remedies. It should be administered in gelatin-capsular form in conjunction with powdered purgatives.

Aspidium is perhaps the most reliable of all the vermifuges, with the exception of pelletierine. For everyday practice it is to be preferred to all other remedies when given in the form of oleoresin. Doses: For adults, 15 to 40 minims; puppies, 5 to 15 minims. The dose of the liquid extract is the same.

Kamala is a very efficient taeniacide with drastic purgative properties. Given in small amount as an adjunct to other taeniacides, particularly to the oleoresin of male fern, it will be found a very valuable remedy. Doses: Adults, 15 to 30 grains; puppies, 3 to 15 grains.

Brayera (U. S. P.), *Cusso* (B. P.), yields kosin or koussin, to which it owes its taeniacidal properties. It is one of the best and safest taeniacides, its action being directly toxic to the worm, but it is too expensive for ordinary practice. The infusion (*Infusum brayerae*, U. S. P.) and fluid extract (*Extractum brayerae fluidum*) are both too bulky and disagreeable for administration to dogs. Kosin may be given in capsules in doses: Adults, 10 to 40 grains; puppies, 10 to 20 grains. The drug usually acts as its own cathartic, but it is better to employ some adjunct for this purpose.

Powdered *areca nut*, when freshly ground, is a very good remedy for tapeworm. When old, it will generally be found inert; consequently, it is best always to purchase the nut and grind or grate on an ordinary nutmeg grater. It is still largely used by British veterinarians and is a favorite with some Americans, but it can not be regarded as being either as effectual or easy of administration as the two preceding drugs. Its effects on puppies are not unattended with danger, on account of its great astringency; but with due regard to subsequent purgation it is a perfectly safe remedy. Mayhew's method of prescribing 1 to 2 grains to every

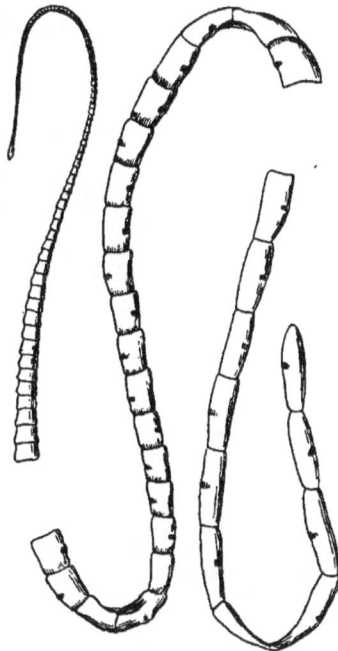

FIG. 95.—An adult Gid Tapeworm (*Taenia coenurus*), natural size. (After Railliet, 1896, p. 244, fig. 135; see also figs. 86c and 88c.) See p. 108.

pound weight of the dog is usually followed, but the smaller quantity will generally suffice, provided the powder is freshly ground. It may be conveniently given in gelatin capsules, accompanied or followed by a purgative.

Turpentine is a powerful remedy against tapeworms, but it is regarded as being somewhat dangerous from its liability to produce strangury and renal inflammation. These effects are said to be less pronounced after large than after small doses; but large doses are more liable to cause gastric and enteric inflammations. It can hardly, therefore, rank with the best remedies. Administer in emulsion with white of an egg, mucilage, milk or oil. Doses: Adults, 10 to 15 minims; puppies, 3 to 10 minims.

Dr. Hoskins has had very satisfactory results with this drug in puppies under 6 months of age and has never noticed any gastric or renal results. In very young puppies he rarely gives over 2 minims, carrying it up to 10 minims, and repeating for two or three days on an empty stomach in the morning, allowing no food for an hour or two after its administration.

The following suggestions as to doses, compiled from various sources, are taken from Curtice (1890):

(1) Allow 2 grains of freshly powdered areca nut for each pound of the dog's weight; administer dose in soup or milk, stirring it well, or by mixing it in butter or molasses. Follow in two hours with a tablespoonful of castor oil for a moderate-sized dog, giving the oil alone or in three times its quantity of milk.

Zürn advises 4 drachms of areca nut for a large dog; 2½ drachms for a medium-sized animal, and 1 drachm for a small dog.

(2) One teaspoonful of turpentine and two tablespoonsful of castor oil given in a cup of milk; the final dose of physic is not given in this case.

(3) Twenty drops of oil of male shield fern, 30 drops of turpentine, and 60 drops of ether. Beat together with one egg and give to the dog in soup.

(4) Hagen advises 80 grains of oxide of copper with 40 grains each of powdered chalk and Armenian bolus; mix with sufficient water to make an adherent mass, and divide into 100 pills. Administer one pill three times daily for ten days in meat or butter.

(5) Röll prescribes the following dose for large dogs; smaller doses should be given in proportion to the size of the dog:

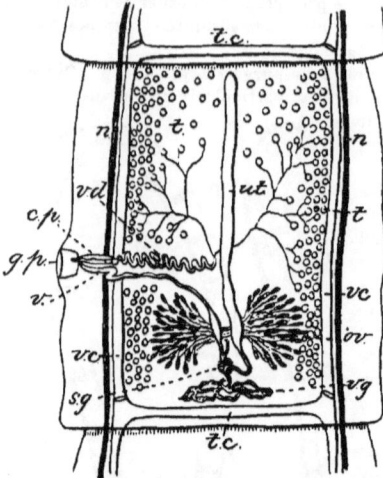

FIG. 96.—Sexually mature segment of the Gid Tapeworm (*Taenia coenurus*): *cp*, cirrus pouch; *gp*, genital pore; *n*, nerve; *ov*, ovary; *sg*, shell gland; *t*, testicles; *tc*, transverse canal; *ut*, uterus; *v*, vagina; *vc*, ventral canal; *vd*, vas deferens; *vg*, vitellogene gland. × 20. (After Deffke, 1891, Pl. I, fig. 3.)

(a) Two drachms each of extract of male fern and of powdered male fern; or—

(b) Decoction of 2½ ounces of pomegranate-root bark in water, reduced to 6 fluid ounces, to which add 1 drachm of extract of male fern. Give in two doses, at intervals of one hour; or—

(c) One-half to 1 ounce of kousso, made into pills, with honey or molasses and a little meal; or—

(d) From 1½ to 2½ drachms of kamala, stirred with honey or water, and given in two doses inside of an hour.

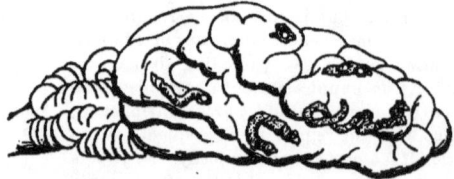

FIG. 97.—Brain of a lamb infested with young Gid Bladder worms (*Coenurus cerebralis*), natural size. (After Leuckart, 1880, p. 456, fig. 206.)

[a, b, and c, should be followed in two hours, with castor oil, but this is not necessary for d.]

After treatment, all the faeces passed during the confinement of the patient should be collected and burned or buried in quicklime.

22. The Gid Bladder Worm (*Coenurus cerebralis*) of Sheep and Calves, and its adult stage, The Gid Tapeworm (*Taenia coenurus*) of Dogs.

[Figs. 87 C, 89 C, 94–100.]

The Gid Bladder Worm is an important and dangerous parasite to the sheep industry, but fortunately it does not seem to be prevalent in this country.

LARVAL STAGE (*Coenurus cerebralis*).

For anatomical characters, compare figs. 97, 99, 100 with key, p. 21.

SYNONYMY.—*Vermis vesicularis socialis* Bloch, 1782; *Taenia vesicularis* Goeze, 1782; *Multiceps* Goeze, 1782; *Hydatigena cerebralis* Batsch, 1786; *Vesicaria socialis* Schrank, (1788); *Taenia cerebralis* (Batsch) Gmelin, 1790; *Polycephalus ovinus* Zeder, 1803; *P. borinus* Zeder, 1803; *Coenurus cerebralis* (Batsch) Rudolphi, 1808.

HOSTS.—Calves, sheep, mufflon, goat, roedeer, reindeer, dromedary, horse. (See pp. 137–143.)

ADULT STAGE (*Taenia coenurus* Küchenmeister, 1853).

For anatomical characters, compare figs. 95 and 96 with key, p. 101.

BIBLIOGRAPHY.—For technical discussion, see especially Deffke (1891).

HOSTS.—Dogs and wolves. (See pp. 137–143.)

Life history.—Starting with the adult worm (fig. 95) in the intestine

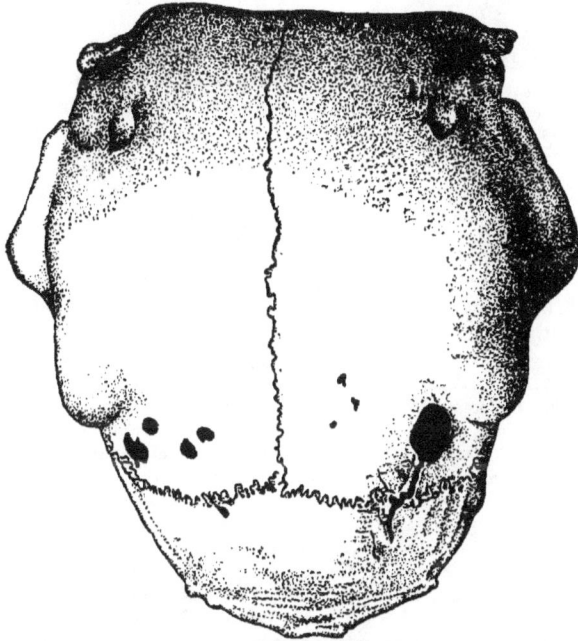

FIG. 98.—Sheep's skull, the hind portion thin and perforated, due to the presence of Gid Bladder worms (*Coenurus cerebralis*). (After Dowitz, 1892, p. 65, fig. 47.)

of the dog, the eggs are scattered on the ground, living three to four weeks in a moist place, and are taken in by the sheep or cattle along with the fodder or water. On becoming free in the intestine, the embryo bores through the intestinal wall and reaches the brain or spinal cord, probably aided in its wanderings by the blood current. Arriving in the brain, the young worm loses its hooks and develops into a cyst (fig. 97), which preserves for some time the power of locomotion and burrows small galleries or canals in the nervous tissue, the canal gradually growing larger as the parasite increases in size. In fourteen to nineteen days after infection, small (0.5 to 1.5 mm.) cysts are found in the brain substance, and similar structures are sometimes found in the muscles, especially of older animals. Those in the muscles generally atrophy in a short time, but those in the brain[1] continue to grow, in twenty-five to forty-five days, causing the symptoms of "gid" or "staggers;" in fifty days they reach the size of a hazelnut and show the anlagen (primordii) of the scolices; in two to three months they complete their development. The heads (figs. 99, 100) form in invaginations generally at one end of the cyst, the invaginations growing in a bunch. If these heads are fed to dogs they develop into adult tapeworms, which produce eggs after about[2] four to eight weeks.

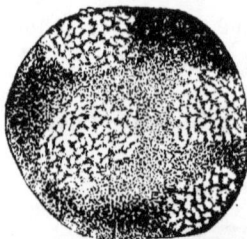

Fig. 99.—An isolated Gid Bladder Worm (*Coenurus cerebralis*), showing the heads. (After Railliet, 1886, p. 245, fig. 137.)

The disease in calves and lambs.—As intimated above, lambs are much more subject to the "gid" than are older animals, a fact which according to some authors finds its explanation in the circumstance that the embryos can not force their way through the tissues of adults, but owing to the more pliable condition of the tissues of young animals they are able to penetrate to the brain without difficulty. In the case of cattle, however, although the disease is more frequent in animals a year or so old, it is not so rare to find 4-year old or 6-year old cows also infested with the parasite.

Three stages of the disease are recognized: (1) The period of infec-

---

[1] While generally found in the nervous centers (brain, more rarely in the spinal cord), it has also been reported once in the connective tissue of sheep (Eichler), once under the skin of a calf (Nathusius), and one extremely doubtful case has been reported to us from Minnesota of its occurrence under the skin of a horse. This latter case has not been examined by the Bureau, but I would suggest that *Taenia serialis* is common in America, and considering the tissue in which this parasite was found, it is not at all improbable that the Minnesota case was one of *Coenurus serialis* (*Taenia serialis*) rather than *C. cerebralis*.

[2] Statements are found in the writings of various authors that *T. coenurus* becomes "ripe," "mature," or "developed" in the dog in "ten days," "three to four" or "six to eight weeks." The expressions "ripe, matured, and developed" are, however, indefinite terms, for in some writings they refer to the stage in which the *genital glands* are active, in other writings they refer to the *gravid* segments. Properly speaking, a segment is *mature* when its sexual glands are active; it is *gravid* or *ovigerous* when it contains embryos. Von Siebold found *gravid* segments thirty-eight days after infection; the strobilae were 16 to 26 inches long, and some segments had already been shed, showing that gravid segments were formed in less than thirty-eight days. Railliet states that according to Leuckart the distal segments "arrive at maturity" (probably meaning "gravid" in this case) after three to four weeks. According to Kreuder the worms develop in dogs "to fully-developed sexually mature tapeworms" in ten days (Zürn).

tion and migration; (2) a period of apparent though not real recovery, and, (3) the climax.

As gid is apparently not prevalent in this country, it is hardly necessary to give a detailed discussion of the symptoms and pathology. The following short account will suffice for the present:

If the parasites are located in the brain, we find the condition known as *cephalic gid*, if in the spinal cord we find *medullary gid*, also known as *lumbar gid* or *hydatic paraplegia*.

In *cephalic gid* there is at first indifference and weakness, an abnormal attitude of the head, which is of an unusually high temperature, and vascular injection of the sclerotica; pressure on the skull causes pain; the most characteristic symptom of the disease is the action of the animal in turning in circles to the right or left, the circles becoming smaller until the patient pivots around on one spot; in some cases it acts as if intoxicated, often stumbles and falls; the eyes are turned in or out, and grinding of the teeth is noticed. The exact position of the parasite determines the particular symptoms.

In *medullary gid* there is a gradual paralysis of the hind legs, paralysis of the rectum and bladder, paleness of the mucous membranes, shedding of the wool, etc.

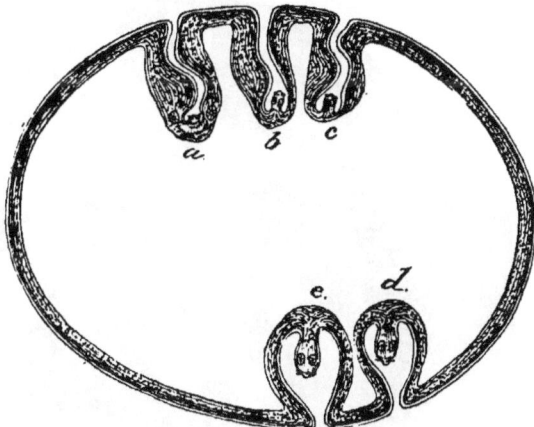

Fig. 100.—Diagrammatic section of a Gid Bladder Worm (*Coenurus cerebralis*): a, normal disposition of scolex; b, c, d, e, diagrammatic drawing to show the homology between cysticercus and coenurus. (After Railliet, 1886, p. 243, fig. 134.)

Cephalic gid should not be mistaken for *vertigo* due to heat, *epilepsy*, *blindness*, or *false gid* due to grubs in the head. *Medullary gid* should not be confounded with the *trembling disease* (the Scotch *louping-ill*) or *lumbar prurigo*.

Treatment.—There is no medical treatment which can be suggested; surgical treatment is sometimes resorted to, but should be performed only by a veterinarian since it is necessary to locate the parasite before operating, and this can be done only by men of experience.

Prevention, however, can and should be practiced by every farmer. Dogs should be kept free from tapeworms. As *Taenia coenurus* develops in the dog in three to eight weeks, the treatment may be repeated two to five weeks after the first dose.

When gid is suspected in a flock of sheep or in cattle, one of the animals should be slaughtered and its skull examined for the parasite. If it is positive that gid is present, it is well to slaughter the affected sheep before the third stage of the disease sets in, as in the third stage they sometimes become very thin. The skulls of "giddy" sheep should not be fed to dogs unless they be first subjected to a long boiling, for if this precaution is not taken the infection will be spread.

The eggs retain their vitality less than twenty-four hours when exposed to an August sun (Leuckart); if kept moist they are alive after three weeks (Gerlach), but after eight weeks are unable to develop further (Leuckart).

It has been impossible for the writer to find any positive evidence of the existence of the Gid Bladder Worm in this country, yet in view of the importations from Europe of sheep and dogs, it is difficult to believe

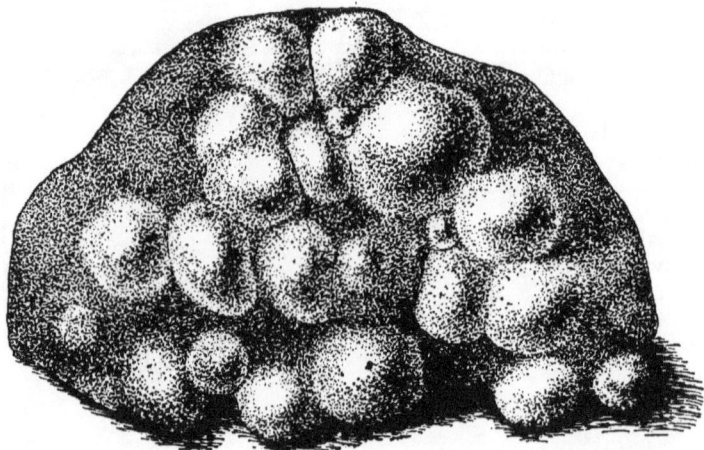

Fig. 101.—Portion of hog's liver infested with Echinococcus hydatid, natural size (original).

that we are entirely free from this parasite. Leidy, in 1856, makes a reference to a parasite "*Coenurus cerebralis* Rud.; in the sheep, *Capra aries*," which he evidently examined, but he gives no details as to where or when the parasite was found. One doubtful case of the presence of adult worm in dogs was recently recorded by Ward, of Nebraska, but I have examined the head of the worm and find it to be a *T. serialis*, see page 101.

ABATTOIR INSPECTION.

So far as infection of man is concerned, the abattoir inspection for *Coenurus cerebralis* is of no importance, for no case of *Taenia coenurus* has ever been recorded in the human species. If the parasite is found in abattoir inspection, care should be taken to dispose of it in such a way (heat) as to render an infection of dogs impossible.

The adult tapeworm in dogs.—Symptoms, etc., see page 101.

23. The Echinococcus Hydatid (*Echinococcus polymorphus*) of Man, Cattle, Sheep, Swine, etc., and its adult stage, The Echinococcus Tapeworm (*Taenia echinococcus*) of Dogs.

[Figs. 101–109.]

A third tapeworm of the dog, the larval form of which develops in cattle, sheep, and swine, is *Taenia echinococcus*. Since this parasite develops its larval stage in man also, and further, since it is the most dangerous animal parasite found in man, it is important to thoroughly understand its life history in order to guard against infection, although it is at present not very common in America.

LARVAL STAGE (*Echinococcus polymorphus*).

For anatomical characters, compare figs. 101, 105, 106–109 with key, p. 21.

SYNONYMY.—*Taenia visceralis socialis granulosa* Goeze, 1782; *Hydatigena granulosa* Batsch, 1786; *Vesicaria granulosa* (Batsch) Schrank, (1788); *Taenia granulosa* (Batsch) Gmelin, 1790; *Polycephalus hominis* Zeder, 1800; *Echinococcus* Rudolphi, 1802; *Polycephalus humanus* Zeder, 1803; *P. granulosus* (Batsch) Zeder, 1803; *P. echinococcus* Zeder, 1803; *Acephalocystis* Laennec, 1804; *Echinococcus granulosus* (Batsch) Rudolphi, 1805; *Hydatis erratica* Blumenbach, 1805; *Acephalocystis humana* Lüdersen, (1808); *A. suilla* Lüdersen, (1808); *Echinococcus hominis* (Zeder) Rudolphi, 1810; *E. simiae* Rudolphi, 1810; *E. veterinorum* Rudolphi, 1810; *Polycephalus granosus* Laennec, 1812; *Acephalocystis oroidea* Laennec, 1812; *A. cystifera* Laennec, 1812; *A. granosa* Laennec, 1812; *A. surculigera* Laennec, 1812; *A. intersecta* Laennec, 1812; *A. ansa* [1] Laennec, 1812; *Echinococcus infusorium* F. S. Leuckart, (1827); *Acephalocystis eremita sterilis* Cruvielhiel, (- ?-); *A. prolifera socialis* Cruvielhiel, (-?-); *A. endogena* Kuhn, (1830); *A. erogena* Kuhn, (1830); *A. granulosa* Chiaje, 1833; *A. communis* Chiaje, 1833; *A. prolifera* Chiaje, 1833; *A. simplex* Goodsir, 1844; (??) *Diskostoma acephalocystis* Goodsir, 1844; (??) *Astoma acephalocystis* Goodsir, 1844; *Echinococcus arietis* E. Blanchard, 1848; *E. giraffae* Gervais, (-?-); *E. polymorphus* Diesing, 1850; *E. pardi* Huxley, (1852); *E. scoliciparens* Küchenmeister, 1855; *E. coenuroides* Küchenmeister, 1855; *E. altriciparens* Küchenmeister, 1855; (?) *Acephalocystis macaci* Cobbold, 1861; (?) *A. ovis tragelaphi* Cobbold, 1861; *Cysticercus echinococcus* (Zeder) Koeberlé, 1861; *Echinococcus cerebri* Spiering, 1862; *E. hepatis* seu *process. vermiformis* Scholler, 1862; *E. hydatidosus* R. Leuckart, 1863; *E. endogena* (Kuhn, 1830) Leuckart, 1863; *E. multilocularis* Leuckart, 1863; *E. lienis* Kehlberg, 1873; *E. pulmonum* Huppert, 1875; *E. multilocularis hepatis* Haffter, 1875; *E. intercranialis* Fricke, 1880; *E. simplex* Leuckart, 1880; *E. racemosus* Leuckart, 1880; *E. multiplex* Stiller, 1882; *E. alveolaris* R. Blanchard, 1886; *E. retroperitonialis* Bitter, 1886; *E. mesenterii* Surmann, 1891; *E. cerebralis* Perroncito, (18—); *E. cysticus* Huber, 1891; *E. unilocularis* Huber, 1896; *E. multilocularis exulcerans* Huber, 1896; *E. osteoklastes* Huber, (?) 1896; *E. subphrenicus* Huber, 1896; "*Echinokokkus*" (!) of Schneidemühl, 1896.
BIBLIOGRAPHY.—For detailed technical discussion of the parasite, see especially Leuckart (1880, I, pp. 732–825); for discussion of hydatid disease in man, see especially Neisser (1877); J. D. Thomas (1884); Davaine (1877, pp. 356–666); for bibliography, see especially (prior to 1864) Diesing (1850, pp. 842–844, and 1864, pp. 395–397); (1861–1880)

[1] *A. plana* Laennec, 1812, a seventh supposed but doubtful variety described by Laennec, has since been determined as a spurious parasite, representing albuminous concretions occasionally found in the wrist, and afterwards described by Dupuytren as *Oculigera carpi*. *A. racemosa* Cloquet, an eighth supposed variety, is another spurious parasite later determined as chorial vesicles.

Taschenberg (1889, pp. 1036–1057); (1877–1890) Huber (1891, pp. 5–39); also Billings, Index Cat. Lib. Surg. Gens. Office, United States Army, 1885, VI, pp. 530–535.
HOSTS.—Man, cattle, sheep, swine, and other animals. (See pp. 137–143.)

ADULT STAGE (*Taenia echinococcus* Siebold, 1853).

For anatomical characters, compare figs. 102–104 with key, p. 101.

SYNONYMY.—"*Taenia cateniformis*" misdet. pro parte Rudolphi, 1808; "*T. cucumerina* Bloch" misdet. pro parte, Diesing, 1850; "*T. serrata*" misdet. Röll, 1852; *T. echinococcus* Siebold, (1853); *T. nana* Beneden, 1858 [nec Siebold, 1852]; *Echinococcifer echinococcus* (Siebold) Weinland, 1858; "*T. echinococca*" of Koeberlé, 1861; *T. (Echinococcifer) echinococcus* of Lenckart, 1863; *T. (Arhynchotaenia) echinococcus* of Diesing, 1861; *T. (Echinococcus) echinococcus* of Railliet, 1886; *T. "echinokokkus"* of Schneidemühl, 1896.
HOSTS.—Dog, dingo, jackal, wolf, cougar (?). (See pp. 137–143.)

Life history.—Starting with the adult tapeworm (fig. 103) in the small intestine of the dog or wolf, the eggs are scattered over the ground and are swallowed by the intermediate host with the fodder or water. Upon arriving in the stomach, the eggshells are destroyed and the six-hooked embryo, which is thus freed, bores its way through the intestinal wall and wanders, actively or passively (that is, carried along by the blood), to various organs of the body, liver, lungs, ovaries, bones, skull, etc., where it develops first into an *acephalocyst,* which may develop further into any of the variations given below in the description of the larval stage.

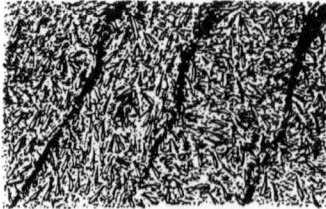

FIG. 102.—Portion of the intestine of a dog infested with the adult Hydatid Tapeworm (*Taenia echinococcus*), natural size. (After Ostertag, 1895, p. 430, fig. 99.)

The heads which are formed, upon being devoured by a dog or wolf, then develop into adult tapeworms.

The larval stage develops rather slowly, and may persist for many years. Thus, cases are on record where the hydatid has existed for 2, 4, 8, 15, 18, and even 30 years in man, very often, however, with fatal results.

Modifications of the hydatid cysts.—The larval stage appears in several different forms, which have been described under various names as representing different species. It is now admitted, however, by nearly all authors, especially by zoologists, that all these forms belong to one species and have been brought about by different modes of growth. Let us assume that a six-hooked embryo has reached the liver, lungs, or some other organ of the secondary host (man, cattle, sheep, etc.).

About four weeks after the infection small cysts, scarcely 1 mm. in diameter, are noticed in the interlobular tissue of the liver, for instance. They consist of an outer cyst, formed by the connective tissue of the host, and an inner solid body, 0.25 to 0.50 mm. in diameter, which represents the young parasite. The six hooks of the embryo have been discarded and the organism consists of an outer transparent capsule—the cuticle—20 to 50 μ in thickness, and a granular content somewhat condensed on the periphery and containing cells which are not distinctly separated from one another. At the end of eight weeks the parasite has doubled in size. The cuticle, which is very elastic, grows thicker and its inner surface is covered by a thin membrane (endocyst, parenchym layer, germinal layer) which represents the condensed granular content; this was at first solid and occupied the entire space inside the cuticle. The endocyst now incloses a cavity containing a clear watery fluid. The parasite continues to grow, the cuticle becomes stratified; the germinal layer shows a histological differentiation into small cells occupying the periphery, large cells on the inside, and granular cells occupying the irregular spaces on the surface. At the end of nineteen weeks the parasite has reached 10 to 12 mm. in diameter; the liquid in the interior contains a number of chemical compositions, the parenchym layer has grown slightly, the cuticle is about 0.2 mm. thick. When the parasite is composed of only these portions, that is, *cuticle, endocyst*, and the contained *liquid*, it represents the form which some authors include under the term *Acephalocystis*. If we imagine all the portions of fig. 105 absent, which are designated by the letters *a* to *z*, the portions *cu* and *pa* being left, we have before us a simple *acephalocyst* (headless echinococcus hydatid). Although the parasite frequently remains in this condition, or rather is found in this condition, the acephalocyst does not represent the final larval stage. Referring to fig. 105, *a*, we see a slight proliferation of the parenchyma. This protuberance grows gradually into the cavity of the hydatid and develops into a brood capsule, *b, c,* the cavity of which is lined by a thin cuticle. The heads of the succeeding generation of tapeworms develop in these brood capsules, but authors are not entirely agreed as to how

Fio. 103.—Adult Hydatid Tapeworm (*Taenia echinococcus*), enlarged. (After Leuckart, 1880, p. 743, fig. 316.)

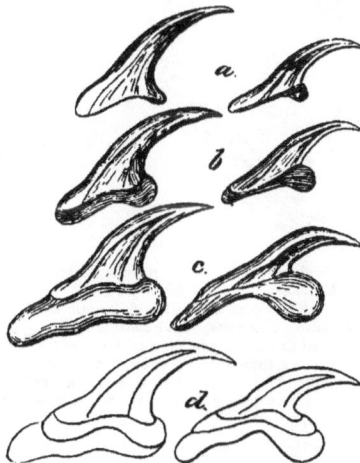

they develop. Thus, Leuckart states that a diverticulum is formed which extends into the cavity of the hydatid cyst, that the head is formed at its base, and the diverticulum then invaginates. The successive stages may be seen in *c, d,* and *e* of fig. 105. Moniez, on the other hand, states that the head develops inside of the brood capsules, passing through the stages *f, g, h,* and *i*; he admits, however, that there is occasionally a diverticulum formed, at the end of which is developed a head, not in the manner described by Leuckart, but in the same manner as if the head had formed inside the brood capsule *f, k.* Whatever may be the mode by which these heads are formed, several (5, 10, 20, or even 34) may develop in one brood capsule. As numerous brood capsules may develop in one hydatid cyst, it is not to be wondered at that many thousand heads are sometimes found in hydatid cysts. Occasionally the brood capsules will be found ruptured, so that the heads extend free into the cavity of the hydatid (*m*), and heads are occasionally found floating free in the liquid of the cyst (*n*). The hydatid, so far as we have traced it (with *cu, pa, a–n*), is

Fio. 104.—Hooks of adult Hydatid Tapeworm: *a*, from a hydatid; *b*, three weeks after feeding to a dog; *c*, from an adult; *d*, combined figures of *a–c*, showing the gradual changes in form. ×600. (After Leuckart, 1880, p. 736, fig. 315.)

a mature larval stage such as is frequently found in animals, and if this cyst is devoured by a dog the separate heads or scolices will develop into adult tapeworms. From this point, or even before it, several modes of development are open for the hydatid: Thus, small centers of growth (*o, p, q, u*) may form in the wall of the parasite. As these growths increase in size a cuticle is formed around them (*p, q*), and they burst through the wall in which they are growing and continue their further development in the same manner as the mother hydatid. If these so-called *daughter cysts* fall into the cavity of the mother cyst, the entire parasitic cyst (mother hydatid + daughter hydatids *r, x*) presents to us the form described as the *endogenous Echinococcus (Acephalocystis endogena* Kuhn, *Echinococcus altricipariens* Küchenmeister, and *E. hydatidosus* Leuckart), found particularly in man, hogs, and

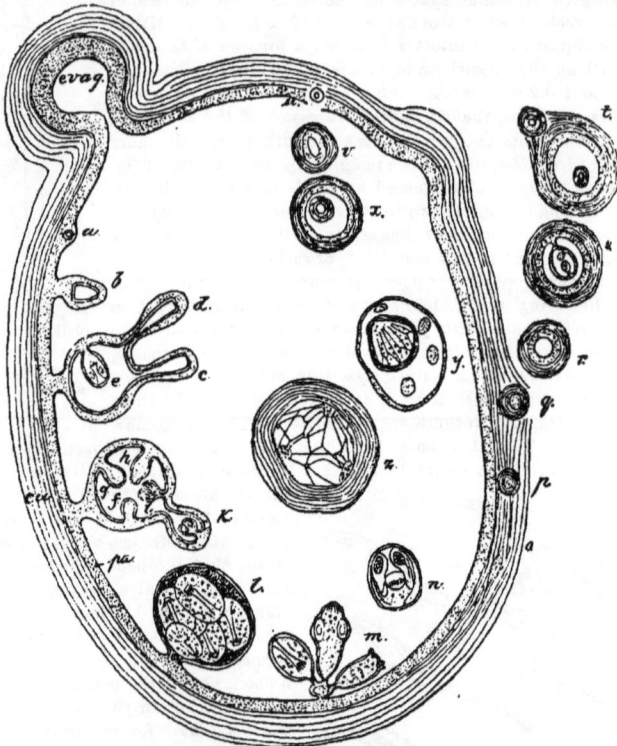

FIG. 105.—Diagram of an Echinococcus hydatid: *cu*, thick external cuticle; *pa*, parenchym (germinal) layer; *c, d, e*, development of the heads, according to Leuckart; *f, g, h, i, k*, development of the heads according to Moniez; *l*, fully developed brood capsule with heads; *m*, the brood capsule has ruptured, and the heads hang into the lumen of the hydatid; *n*, liberated head floating in the hydatid; *o, p, q, r, s*, mode of formation of secondary exogenous daughter cyst; *t*, daughter cyst with one endogenous and one exogenous granddaughter cyst; *u, v, x*, formation of endogenous cyst, after Kuhn and Davaine; *y, z*, formation of endogenous daughter cysts, after Naunyn and Leuckart: *y*, at the expense of a head; *z*, from a brood capsule; *evag.*, constricted portion of the mother cyst. (After R. Blanchard, 1886, p. 426, fig. 257, slightly modified.)

horses. The growth does not necessarily stop with the daughter cysts, but a third generation of cysts (*granddaughter cysts*) may form in the same manner inside of the daughter cysts, as shown in *x*. The brood capsules of the mother cyst, or even the separate scoleces, may, according to certain authors, fall into the cavity and develop into daughter cysts. If the daughter cysts continue their growth outside of the mother cyst, as shown in *q, r, s, t*, we have the form described as the *exogenous Echinococcus (Acephalocystis exogena* Kuhn, *Echinococcus scolicipariens* Küchenmeister, *E. simplex*, and *E. granulosus* Leuckart). It will be at once seen that it is sometimes difficult to decide whether the parasite is an exogenous echinococcus or whether the cysts *s* and *t* have developed from six-hooked embryos.

In the case of *endogenous echinococcus* it would not be at all strange if we found the scolices free in the liquid *n* or a ruptured brood capsule *m*, caused by contact of the brood capsule with the daughter hydatids.

Generally, the hydatids are more or less round, but frequently diverticula (*evag.*) are noticed in the walls; for the cyst will naturally develop in the direction of the least pressure, and if this pressure is least at one particular portion of the cyst a diverticulum will naturally form at that point. This growth of diverticula leads us to the consideration of the form of hydatid known as *Echinococcus racemosus* and still another form not distinctly separated from *E. racemosus*, that is, *E. multilocularis* (*E. alveolaris*, the *"tumeur hydatique alveolaire"* of Carriére).

E. racemosus Leuckart, the grape Echinococcus (fig. 106), is composed of a number of cysts more or less intimately connected with each other, so as to give the appearance of a bunch of grapes or of fish spawn. It is difficult to distinguish in some cases whether the parasitic growth represents a heavy infection of small hydatids, each of which has grown from a six-hooked embryo, or whether all the cysts have arisen by budding from a single cyst. Cases of this kind have been reported in cattle by Kuhn and others.

FIG. 106.—A racemose Echinococcus, natural size. (After Leuckart, 1880, p. 795, fig. 334.)

E. multilocularis,[1] as stated, is a form of growth which is not distinctly separated from *E. racemosus;* in fact, the two may easily be classed together, as Leuckart suggests. *E. multilocularis* s. st. represents a group of small hydatids (figs. 107–109) lying close together, in many cases connected in a common stroma. This variety is found chiefly in Switzerland and Germany, where about 70 cases have been reported in the liver of man and a number of cases in cattle.

If a section is made of the parasitic growth we find numerous small caverns of irregular shape, containing a rather transparent gelatinous substance and embedded in a common substance or stroma of connective tissue, in which blood vessels and gall ducts are occasionally seen. The liver cells, however, are entirely atrophied. For many years the nature of these parasitic growths was misunderstood and they were diagnosed as colloid cancers until Virchow (1856) discovered that they were hydatids.

Four other terms which have been applied to the hydatids also need a word of explanation. Rudolphi made use of the terms *E. hominis, E. simiae,* and *E. veterinorum* to designate the echinococcus of man, apes, and other animals, respectively, supposing that they belonged to three separate species. Diesing, however, maintained that all three forms represent the larval stage of one species and introduced the name *E. polymorphus* to designate the larval parasite, a name which zoologists now quite generally accept.

HYDATID DISEASE IN VARIOUS ANIMALS.

FIG. 107.—Section through a multilocular Echinococcus. ×30. (After Leuckart, 1880, p. 796, fig. 335.)

The disease caused by the larval stage of this parasite is known as Hydatid, or Echinococcus, disease. In general terms, the hydatid may occur in any organ of the body, but is most commonly met with in the liver or lungs. The symptoms will of course vary according to the location of the parasite.

[1] Several authors, more particularly Müller and Mangold, consider that this form represents a distinct species. The ordinary adult is said to have plumper hooks, while the eggs are not collected in "egg balls." The adult of *E. multilocularis* is said to have more slender hooks and its eggs are described as collected in "egg balls."

It is not at all rare that the hydatids are not known to be present until discovered by post-mortem examination; they may however become very dangerous because of their situation, their volume, and the pressure they exert. When they occupy an important organ, when they reach a large size, and when the walls of the cysts become osseous or cartilaginous; or when numerous, they may cause serious trouble or death; they are frequently fatal when after bursting they are discharged through an organ communicating with the exterior, symptoms persisting and increasing, the expelled matter having a gangrenous odor, or when they discharge into a serous cavity or into a large blood vessel.

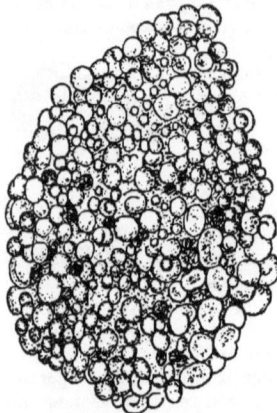

FIG. 108.—A multilocular Echinococcus from the liver of a steer, natural size. (After Ostertag, 1895, p. 427, fig. 94.)

Hydatid disease in cattle.—Three cases are known where cows died suddenly which had hydatids in the heart; Fuisen is authority for the statement that the hydatids of cattle are short lived, and show a great tendency to degenerate and become calcified. (See also under "Abattoir inspection," p. 121.)

Symptoms.—The parasites are generally found in the liver and lungs, seldom in the heart. When in the heart symptoms are not generally exhibited unless the cyst breaks through the muscular wall and hangs into the cavities of the heart, or when the cyst discharges; in these cases apoplexy is generally the result. It is scarcely possible to diagnose echinococcus of the spleen. In echinococcus of the lungs a slight cough is first noticed, which increases according to the degree of infection and the size of the parasites, occurring at times every five or ten minutes. This cough is absent when the liver instead of the lungs is particularly infected. Respiration increases to 80 or 84 per minute. Inspiration is broken. Fever is at first absent; pulse about 70 to 85; milk secretion is lessened, appetite normal

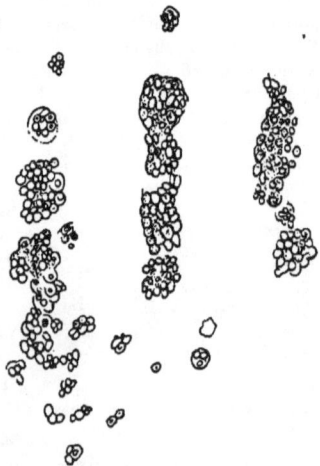

FIG. 109.—A multilocular Echinococcus from the pleura of a hog, natural size. (After Ostertag, 1895, p. 428, fig. 97.)

except toward the end of the disease, when the hide becomes bound, hair becomes stiff and dry. Pressure on the right side of the region of the four last ribs causes the animals to show signs of pain, and there is a dull percussion sound similar to, though not so deep as that in pleuro-pneumonia, covering small areas or the entire breast and the region of the right lobe of the liver. Placing the ear on the chest one hears a heavy harsh breathing mixed with other sounds, whistling, rattling, or at the moment of inspiration, an exceedingly characteristic tone which Harms has named "Guurksen" (cloc-cloc of Hartenstein), and which one hears when he presses and shakes bladders filled with liquid. In liver echinococcus the labored breathing is generally absent, but digestive troubles are present, appetite and rumination become irregular; intestinal catarrh, indigestion; a yellowish color of the eyes are noticed. (Abstracts from Zürn (1882, p. 136) and others; Harm's Die Echinococcus-Krankheit des Rindes, 1870, is not accessible here.)

A rectal exploration occasionally shows an enormously enlarged liver, and thus directs suspicion to the disease.

In ante-mortem examinations hydatid disease of the lungs in cattle

Fig. 110.—Lymphatics of a steer infested with the so-called "Tongue worm" (*Linguatula rhinaria.*
(After Ostertag, 1895, p. 434, fig. 102.)

may be mistaken for pleuro-pneumonia, but in the latter disease the sounds upon percussion are deeper and duller than in the former disease. It will be recalled that contagious pleuro-pneumonia is not found in the United States.

For differential diagnosis in post-mortem examinations, see page 121.

Hydatid disease in sheep.—Very little is written upon this subject, but from the data published the symptoms shown by sheep are as vague and indefinite as those exhibited by cattle.

Feebleness, dullness, and indifference, though these may not be very marked, except at the last stages of the malady, when the animal is cachectic. There are frequent tympanites, and pruritus at various points; the wool is dry and brittle and easily pulled out, and, in general, the symptoms are confounded with those of *fascioliasis (distomatosis)*. (Neumann.)

Hydatid disease in swine.—No particular symptoms have been described for hydatid disease in swine.

Pathology.—The pathological lesions naturally vary according to the organs in which the parasites are situated. There may be an enormous increase in the size and weight of the lungs or liver. The normal weight of the liver of an ox is about 5 kilograms (11 pounds), but hydatid livers have been recorded which weighed 50 kilograms (110 pounds), 130 pounds, 145 to 146 pounds, and even 158 pounds. A pig's liver weighs on an average about 2 kilograms (4.4 pounds) when normal, but hydatid livers have been recorded weighing 50 to 100 pounds. A steer's lungs, normal weight 6 pounds, may increase to 40 or 54 pounds.

The increase in size of these organs by the growth of the parasites naturally causes a displacement of other organs; the curvature of the diaphragm is changed; the intestines may be compressed and constricted; adhesions may form; the surface of the organs containing the parasites naturally assumes an abnormal outline, bulging out at points corresponding to the hydatids.

The hydatids themselves cause an atrophy of the specific tissue of the organ, the connective tissue of which proliferates and forms a capsule immediately surrounding the parasite; the surface of this capsule is smooth and glistening, and entirely separated from the cuticle of the cyst, so that with care the parasite may be freed without injury; the capsule grows in thickness from 1 to 10 mm. After a time the cysts may undergo degeneration; the entire body may be replaced by a caseous or gelatinous amorphic mass in which hooks or remnants of the cuticle may be found. The multilocular echinococcus presents an appearance differing from that of the ordinary form and resembling a cauliflower to some extent.

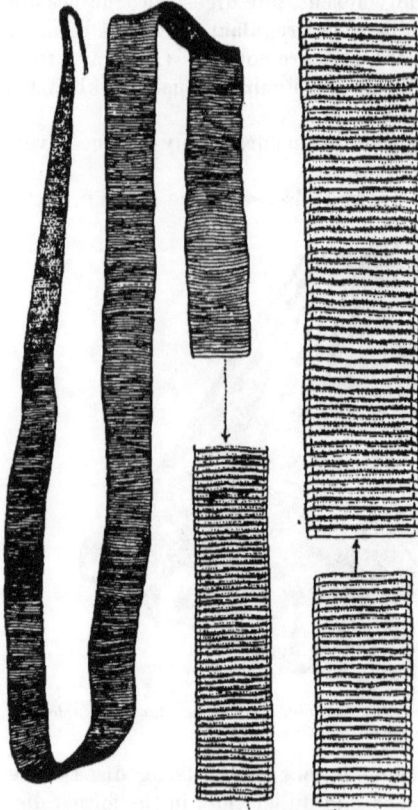

Fig. 111.—Portions of an adult Flat Moniezia (*Moniezia planissima*). (After Stiles & Hassall, 1893, Pl. I, fig. 1). See p. 127.

Differential diagnosis.—In post-mortem examination, hydatid disease, especially of the lungs, may occasionally be mistaken for tuberculosis, more particularly when the hydatids are very young and numerous, or degenerated; in tuberculosis, however, (1) the neighboring[1] lymphatics will generally be involved, which is not the case in hydatid disease, while (2) in hydatid disease the parasite is generally easily separated from its surrounding capsule, (3) the elastic cuticular membrane is lamellated, and (4) a microscopic examination will in some cases show the hooks of the heads. See also p. 79.

Treatment.—It is useless to waste time in trying to treat a domesticated animal in which echinococcus is suspected unless the animal is an especially valuable one, and unless the parasite is located in an organ which can be reached by surgical interference.

A number of methods for treatment in man have been suggested from time to time, but surgical interference is the only one which has been followed by satisfactory results. For a discussion of this subject with citation of cases, see Davaine (1879, pp. 592-663).

Prevention.—Keep dogs away from slaughterhouses. This will prevent their becoming infected with the tapeworms, and thus prevent their transmitting the parasite to man and animals. Stray dogs should be killed; all other dogs should be looked upon as suspicious characters, and should not be accorded the privileges of human beings.

FIG. 112.—Three views of heads of the Flat Moniezia (*Moniezia planissima*). ×17. (After Stiles & Hassall, 1893, Pl. I, figs. 2-2b.) See p. 127.

ABATTOIR INSPECTION.

Organs infested with echinococcus are not directly harmful to man as food, since the parasite will not come to maturity in man's intestine, and there is no objection to placing these organs on the market after the portion containing the parasite has been removed. Removing and destroying the infected portions are precautions which should always be taken in order to prevent the possibility of the further infection of dogs.

The abattoir is the proper place to attack this disease, and a careful and persistent destruction of the larval stage found in meat inspection must finally result in lessening and even exterminating the disease. Heat should be used in destroying the parasite.

Frequency of the hydatid in various animals.—The frequency of hydatid varies greatly in different countries. According to statistics thus far published the parasite appears to be most frequent in Iceland, India, Eastern Siberia, and Australia; it is more common in Mecklenburg

---

[1] An infection of the lymphatics (fig. 110) with the so-called "Tongue worm" (*Linguatula rhinaria*) should not be mistaken for tuberculosis.

than in any other part of continental Europe. The United States seems to be comparatively free from hydatid infection, although the disease is apparently on the increase.

United States.—I have seen cases of hydatids in this country in cattle, hogs, the camel, and man, but as yet have seen no cases in sheep. Wheeler records 117 cases of liver echinococcus in 2,000 hogs examined at New Orleans; the cases in domesticated animals which I have examined came from the District of Columbia, Missouri, and Nebraska; Welch records it for Maryland and several of the Bureau inspectors report it for various abattoirs. (For the cases in man, see p. 124.)

Iceland.—The statistics for Iceland are not altogether satisfactory, but it is alleged that in some districts every sheep of three years old is infested, while it is an exception to find a cow ten years old which is free from this parasite; in some districts it is estimated that about one-third of the sheep are infested; one author estimates that one-fourth are infested.

India.—Seventy per cent of the cattle are infested (Neumann).

Germany.—The statistics for Germany are more detailed than for any other country; it must, however, be borne in mind that while some

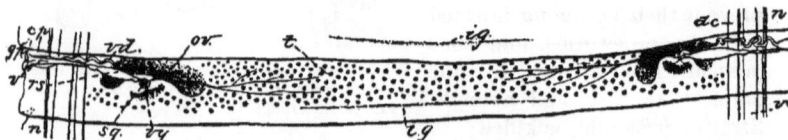

FIG. 113.—Dorsal view of sexually mature segment of the Flat Moniezia (*Moniezia planissima*): *cp*, cirrus pouch; *dc*, dorsal canal; *gp*, genital pores; *ig*, interproglottidal glands; *n*, nerve; *ov*, ovary; *rs*, receptaculum seminis; *sg*, shell gland; *t*, testicles; *v*, vagina; *vc*, ventral canal; *vd*, vas deferens; *vg*, vitellogene gland. Enlarged. (After Stiles & Hassall, 1893, Pl. II, fig. 4.)

German statistics include the whole number of animals slaughtered and the entire number of cases found, other statistics leave out the calves and omit from the list those cases of light infection in which the portion containing the parasite could be excised and the rest of the organ placed upon the market. The following statistics are compiled from various sources:

Peiper (1894) takes the statistics of 52 slaughterhouses in various parts of Germany and concludes that 10.39 per cent of the cattle, 9.83 per cent of the sheep, and 6.47 per cent of the hogs harbor hydatids; the average for Greifswald, Wolgast, Anklam, Demmin, and Swinemünde (Vorpommern) was: Cattle, 37.73 per cent; sheep, 27.1 per cent; hogs, 12.8 per cent; for Greifswald alone, cattle, 64.58 per cent; sheep, 51.02 per cent; hogs, 4.93 per cent.

Mecklenburg.—About half of the animals are infested (Sahlmann). Cows infested to 25 per cent, sheep 15 per cent, hogs 5 per cent (Metelmann).

Stettin.—Cows (293:1425), 7.1 per cent; hogs (1238:16829), 7.3 per cent; sheep (3807:14717), 25.8 per cent (Olt).

Leipzig (one year).—Sheep (591:4515), 13.09 per cent; native hogs (196:5166), 3.79 per cent; Hungarian hogs (181:843), 27.47 per cent. In native hogs the liver (3.81 per cent) was more frequently infected than the lungs (0.26 per cent); in Hungarian hogs liver, 12.03 per cent; lungs, 14.79 per cent; in sheep the lungs (12.71 per cent) were more frequently infested than the liver (3.73 per cent). (Mejer.)

The *Berlin* statistics (quoted from Braun, 1895, who takes them from the Berichte über die städtische Fleischbeschau in Berlin) are especially instructive; they are here reduced to percentages in order to bring out the results more prominently:

Number of organs of cattle, sheep, and hogs condemned for hydatids from 1888 to 1893.

CATTLE (CALVES NOT INCLUDED).

Year.	Number examined.	Condemned for hydatids.			
		Lungs.		Livers.	
		Number.	Per cent.	Number.	Per cent.
1888–89	141,814	6,578	4.6	2,668	1.8
1889–90	154,218	7,266	4.7	2,418	1.5
1890–91	124,593	5,792	4.6	1,938	1.5
1891–92	136,368	4,497	3.2	1,721	1.2
1892–93	142,874	2,563	1.7	739	.5

SHEEP.

1888–89	338,708	5,041	1.4	3,363	0.9
1889–90	430,362	5,479	1.2	2,742	.6
1890–91	371,943	4,595	1.2	2,059	.5
1891–92	367,933	4,435	1.2	1,669	.4
1892–93	355,949	3,331	.9	1,161	.3

HOGS.

1888–89	479,124	5,910	1.2	5,285	1.1
1889–90	442,115	6,523	1.4	5,078	1.1
1890–91	472,859	5,083	1.07	3,735	.07
1891–92	530,551	6,037	1.1	4,374	.08
1892–93	518,073	6,785	1.3	4,312	.08

1893–94 in all 13,424 lungs and 6,283 livers.. (Berichte ü. d. städtische Vieh- u. Schlachthbf.)

These statistics show that from 1888–89 to 1892–93 there has been a reduction in the number of organs condemned for hydatids both in cattle and sheep, which must be attributed to the system of abattoir inspection, and which must necessarily result in a corresponding decrease in hydatid disease in man. This reduction is not so apparent among hogs, but it must not be forgotten that Berlin slaughters large numbers of hogs imported from districts in which the slaughterhouse inspection is exceedingly superficial. We saw above that some German importations of hogs from Russian Poland, Bohemia, etc., were infected with *Cysticercus cellulosae* to 50 per cent, and hogs which are kept in such a manner as to allow this infection will certainly also bring up the German statistics of hydatids. I am strongly inclined to give much greater importance to the Berlin statistics than appears from the percentages of infection among the hogs.

THE ADULT TAPEWORM IN DOGS.

(See p. 101.) It seems to me entirely impracticable to attempt to guard against hydatid disease by trying to definitely diagnose the presence of the adult worms in dogs. If, however, the worm is found in dogs, the latter should be killed and burned. The hydatid is altogether too dangerous a parasite in man to warrant a person's treating a dog which harbors *Taenia echinococcus.*

A decision of the "Professoren-Kollegium des Tierarznei-Instituts zu Brüssel," though amusing to Americans, is of great importance to any country in which canine flesh is used as food; that is, that the oesophagus, stomach, and intestine of all slaughtered dogs are to be excluded from the market.

HYDATID DISEASE IN MAN.

It is important to consider this subject in this connection in order to insist upon the necessity of destroying hydatids found at abattoirs. Hydatid disease is the most fatal zoo-parasitic disease which affects man, "50 per cent of the cases dying within five years after infection," but its occurrence in man is fortunately comparatively rare in this country. One of the volunteer assistants in the Bureau, Dr. H. O. Sommer (1895-96), has recently compiled 100 cases which have been found in the United States. Many of the cases were among foreigners, and some of these were certainly infested before coming to this country.

The 100 cases in the United States were distributed as follows:

BY NATIONALITY.

Nationality.	Cases.	Nationality.	Cases.	Nationality.	Cases.
Azorian	1	Italian	5	Welsh	1
"Colored"	1	Japanese	1	"White"	4
English	5	Mexican	1	Unstated	51
"Foreigners"	2	"Mulatto"	2		
French	2	Negro	2	Total	100
German	15	Pole	1		
Irish	2	Swede	1		

By sex: Males, 47; females, 28; unstated, 25.

BY STATES.

State.	Cases.	State.	Cases.	State.	Cases.
Alabama	2	Louisiana	1 (or 2?)	Pennsylvania	10
California	1	Massachusetts	5 (or 6?)	Tennessee	1
Connecticut	1	Missouri	7	Texas	1 (+?)
District of Columbia	4	Michigan	1	Vermont	1
Illinois	3	New Jersey	1	Virginia	2
Indiana	1	New York	33	Washington	1
Kentucky	2	Ohio	7	Unstated	15

Of 981 cases from various parts of the world, the greatest number occurred in persons between 21 and 40 years of age, as shown by the following classification by ages:

Years.	Cases.	Years.	Cases.
0 to 10	54	51 to 60	82
11 to 20	152	61 to 70	36
21 to 30	274	71 to 81	18
31 to 40	225	Over 80	2
41 to 50	138		

In man the organs most frequently infested are the liver, lungs, kidneys, and cranial cavity. Thus, of 1,806 cases of organ infections, the liver was infested in 1,011 cases, lungs in 147, kidneys in 126, and cranial cavity in 95.

Hydatid disease is especially common in Iceland and Australia. For Iceland the statistics are very contradictory, some authors estimating that 2 per cent, others 16⅔ per cent (probably exaggerated), of the inhabitants are infested. Three thousand cases are reported for Australia between 1861 and 1882.

In central Europe the hydatid is found on an average once in every 130 post-mortems. The frequency varies in different localities, Mecklenburg and Pomerania leading the list. Ostertag gives the following statistics:

Locality.	Cases.	Post-mortems.	Per cent.
Rostock	25	1,025	2.43
Breslau	20	1,360	1.47
Berlin	33	4,470	0.76
Göttingen	3	639	.46
Dresden	7	2,002	.34
Vienna	3	1,229	.24
Prague	3	1,287	.23
Erlangen	2	1,812	.11

Peiper collected 150 cases for Vorpommern from 1860 to 1890; in postmortems at the Pathological Institute of Greifswald the percentage was 1.9.

Prevention of the disease in man.—The disease may be prevented in three ways—

(1) By recalling that the dog is not a human being and should not be treated as one. Too intimate association with dogs is sure to breed the disease in man.

(2) By preventing infection among dogs. This can be done by keeping dogs away from slaughterhouses, and by the destruction (by heat) of all hydatids found in slaughtered animals. The slaughterhouse is the best place to institute measures against hydatid disease in man.

(3) By killing all stray and ownerless dogs.

Adult Tapeworms of Cattle and Sheep (Subfamily *Anoplocephalinae*).

Adult tapeworms are more or less frequently found in the intestines of cattle and sheep, more rarely in the bile duct of sheep. As stated on page 68, they all belong to the subfamily *Anoplocephalinae;* they are very closely related to the tapeworms of horses, hares, and rabbits, and yet are entirely distinct from these forms. ·

Owing to many misidentifications of tapeworms which have been published, and to the meagre descriptions of some of the species, it is impossible to state exactly how many different forms actually occur in cattle and sheep, but we are now in a position to clearly define the most common forms which occur, especially those which are found in this country, and to suppress some of the worms which have been described as distinct species of parasites in these animals, but which in reality are identical with forms previously described under other names, or are parasites erroneously attributed to these hosts.

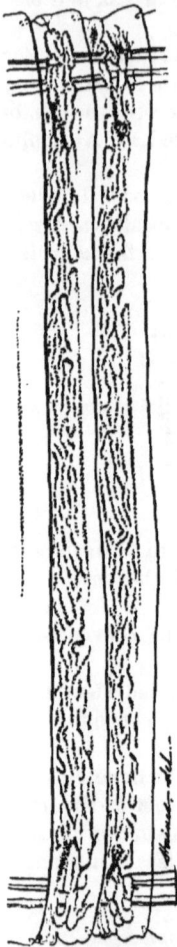

Cattle.—Eight different species of tapeworm have been reported from cattle, but in all probability only four of them are found in this host; these four species all belong to the genus *Moniezia*, and two of them, namely, *Moniezia planissima* and *M. expansa*, are found in this country.

This Bureau has knowledge of only two adult tapeworms in American cattle, but I have examined specimens of three other species, namely, *Moniezia alba*, *M. Benedeni*, and *M. denticulata*, preserved in various European collections and bearing the label that they were taken from cattle. Of these three forms, *M. denticulata* (=*Cittotaenia denticulata*) is unquestionably a parasite of rabbits instead of cattle (Stiles & Hassall, 1896), and an error must have been made in the original label; *M. alba* and *M. Benedeni* are evidently legitimate cattle parasites. Rivolta (1878) states that he examined a tapeworm collected by Perroncito from the ox which he (Rivolta) considered identical with a worm he at first labeled "*Taenia denticulata* (?)" and which he later described as *Taenia orilla* (=*Thysanosoma Giardi*). Perroncito has, however, recently stated to Lungewitz (1895, p. 6) that he found this worm only in sheep. *Thysanosoma Giardi* is accordingly not yet established as a bovine parasite. Von Linstow (1889, p. 20) includes two other tapeworms, namely, *Stilesia centripunctata* and *S. globipunctata*, as parasites of cattle, but I am unable to find the authority for this statement.

Sheep.—A large number of tapeworms have been described or recorded as parasites of sheep, but the number of species must be considerably reduced, for some of the forms described as distinct species are identical with forms previously described under other names, while other forms were misdetermined. Four species, namely, *Moniezia planissima*, *M. expansa*, *M. trigonophora*, and *Thysanosoma actinioides*, are known to occur in American sheep.

Several other forms, namely, *Moniezia alba*, *M. Benedeni*, *M. Neumanni*, *M. nullicollis*, *M. Vogti*, *Thysanosoma Giardi*, *Stilesia centripunctata*, and *S. globipunctata*, occur in sheep in other countries. *Moniezia denticulata* (=*Cittotaenia denticulata*) of the rabbit has erroneously been reported from sheep in Europe.

Fig. 114.—Dorsal view of gravid segments of the Flat Moniezia (*Moniezia planissima*), showing the uterus, enlarged. (After Stiles & Hassall, 1893, Pl. II, fig. 5.)

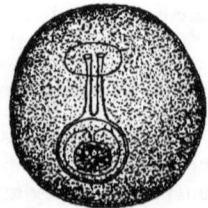

Fig. 115.—Egg of the Flat Moniezia (*Moniezia planissima*), greatly enlarged. (After Stiles & Hassall, 1893, Pl. II, fig. 6.)

Swine.—No species of adult tapeworm is positively known to be a normal parasite in swine, but Cholodkowsky (1894, pp. 552–554) records specimens of *Thysanosoma Giardi* said to have been taken from hogs in Russia, and Detmers (1879), and Stiles (1895, pp. 220–222) have recorded three cases of other forms alleged to have occurred in this country.

The three genera of adult tapeworms for us to consider in connection with cattle, sheep, and swine are *Moniezia*, *Thysanosoma*, and *Stilesia*.

For a technical discussion of these genera and their species, with bibliographies, see Stiles & Hassall (1893) and Stiles (1896). For convenience of discussion, all of the forms will be treated together. For anatomical characters, compare figs. 111–124 with the key, page 21.

GENUS MONIEZIA.

It is often quite difficult to distinguish between the different forms, as the specific characters must to a great extent be taken from the internal anatomy, and it is therefore necessary to make a microscopic examination of one or more specimens which have been artificially stained. In many cases, however, these characters may be recognized if a fresh worm is allowed to macerate one or two days in water; then by pressing some of the segments between two pieces of glass and holding them to the light some of the internal anatomy can be recognized.

24. The White Moniezia (*Moniezia alba*) of Cattle and Sheep.

SYNONYMY.—*Taenia alba* Perroncito, 1879; *Moniezia alba* (Perroncito) R. Blanchard, 1891; (?) *M. alba* var. *dubia* Moniez, 1891.

This tapeworm has been recorded for France, Italy, and Algeria, but not as yet for this country. Poorly preserved specimens of *M. planissima* resemble this form in that the interproglottidal glands can not be seen distinctly. This renders it possible that *M. alba* is simply a poorly preserved *M. planissima*—a point which can not, however, be demonstrated by a comparison of the original types; on this account, it is necessary to retain both species.

25. Vogt's Moniezia (*Moniezia Vogti*) of Sheep.

SYNONYMY.—*Taenia Vogti* Moniez, 1879; *Anoplocephala Vogti* (Moniez) Moniez, 1891; *Moniezia Vogti* (Moniez) Stiles & Hassall, 1896.

Very little is known about this supposed species, which may be a distinct form or may be a dwarfed specimen or some other species. It has been found once in France and once in England, but is not yet recorded for America.

26. The Flat Moniezia (*Moniezia planissima*) of Cattle and Sheep.

[Figs. 111–115.]

SYNONYMY.—*Moniezia planissima* Stiles & Hassall, 1892; *Taenia* (*Moniezia*) *planissima* (Stiles & Hassall) Braun, 1895; *T. expansa* pro parte of various authors.

This seems to be the most common adult tapeworm in American cattle; it also

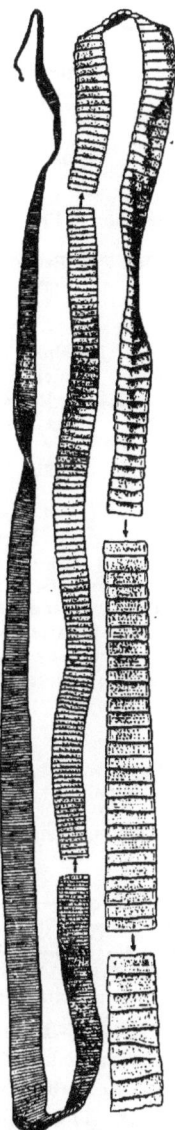

FIG. 116.—Portions of an adult specimen of the Broad Moniezia (*Moniezia expansa*), natural size. (After Stiles & Hassall, 1893, Pl. VI, fig. 1.)

occurs in American sheep, but is apparently not so common in this host. It is found also in France, Germany, and Italy.

27. Van Beneden's Moniezia (*Moniezia Benedeni*) of Cattle and Sheep.

SYNONYMY.—*Taenia Benedeni* Moniez, 1879; *Moniezia Benedeni* (Moniez) R. Blanchard, 1891.

This worm was recorded once for sheep in France and once for cattle in Austria.

28. Neumann's Moniezia (*Moniezia Neumanni*) of Sheep.

This worm was described by Moniez in 1891, and has been recorded only once. It was found in France.

29. The Broad Moniezia (*Moniezia expansa*) of Cattle, Sheep, Goats, etc.

[Figs. 116–119.]

SYNONYMY.—? *Taenia ovina* Goeze, 1782; ? *Halysis ovina* (Goeze) Zeder, 1803; ? *T. expansa* Rudolphi, 1805 (nomen nudum); *T. expansa* Rudolphi, 1810; *Alyselminthus expansus* (Rudolphi) Blainville, 1828; *Moniezia expansa* (Rudolphi) R. Blanchard, 1891; *Taenia* (*Moniezia*) *expansa* of Braun, 1895.

This worm is quite common in America and Europe, both in cattle and sheep.

30. The Triangle Moniezia (*Moniezia trigonophora*) of Sheep.

[Figs. 120–121.]

SYNONYMY.—*Moniezia trigonophora* Stiles & Hassall, 1893; *Taenia* (*Moniezia*) *trigonophora* (Stiles & Hassall) Braun, 1895. Also *T. expansa* and *T. Benedeni* pro parte of some authors.

This is rather a common parasite of American sheep, and is also found in France. It takes its name from the triangular arrangement of the testicles. I have seen one serious outbreak of disease in sheep due in part to this parasite and in part to the twisted wireworm (*Strongylus contortus*) of the stomach.

Genus THYSANOSOMA.

Represented by one species in North America and South America and one species in Europe.

31. The Fringed Tapeworm (*Thysanosoma actinioides*) of Sheep, Deer, etc.

[Figs. 122–124.]

SYNONYMY.—*Thysanosoma actinioides* Diesing, 1835; *Taenia fimbriata* Diesing, 1850 [nec Batsch, 1786]; "*Taenia expansa*" misdet. pro parte, of Faville, 1885; *Moniezia fimbriata* (Diesing) Moniez, 1891.

The Fringed Tapeworm is found in North America and South America, and forms at times a veritable scourge to the sheep industry of the Western plains.

Disease.—The disease in sheep caused by the Fringed Tapeworm has been studied in detail by Curtice (1890, pp. 91–109), who considers that next to scab it is the most important sheep disease of the Western plains. The financial loss it causes is quite extensive, and results from the failure of the lambs to fatten, the lessening of the wool, and the weakening of the animals so that they can not withstand the cold winter weather. The parasites develop slowly, and are present in considerable numbers before their presence is suspected. Toward September the lambs fail to grow as they should; in November the symptoms are quite marked. First, the worms produce a local irritation of the intestine, which finally develops into a chronic catarrhal inflammation; their presence in the gall ducts produces similar results and obstructs the flow of bile; infected lambs are large headed, undersized, and hidebound; their gait is rheumatic and they appear more foolish than the other sheep, standing oftener to stamp at the sheep dogs or herders, and lagging behind the flock when driven; the general symptoms are those of malnutrition, and Curtice considers them nearly identical with the symptoms of the loco disease; in fact, he states that it is extremely difficult to distinguish between the two diseases, and believes that the fact that the worms "may tend to produce depraved appetites and a morbid craze for a particular food is also reason for suspecting that the loco disease may depend on the tapeworm disease." General systematic disturbances result from malnutrition; the usual fat is absent; serous effusions are noticed in the body cavities, serous infiltration in the connective tissue.

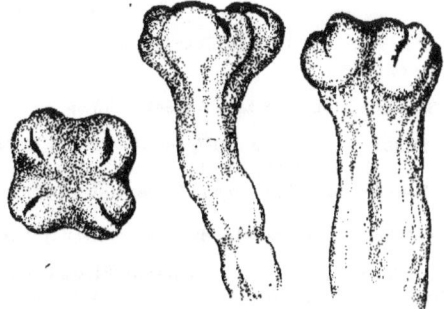

FIG. 117.—Three views of the head of the Broad Moniezia (*Moniezia expansa*). ×17. (After Stiles, 1893, Pl. V, figs. 1–1b.) See p. 128.

Treatment.—Curtice found that powdered preparations of pumpkin seed, pomegranate-root bark, cusso, kamala, male fern, and worm seed were of no avail, a failure due, he maintains, to the anatomical structure of the sheep's stomach and method of administration; no medicine could be used to dislodge the parasites from the gall ducts.

Personally, I have never treated sheep for the Fringed Tapeworm, but I would suggest the advisability of trying the method described on pp. 133–135.

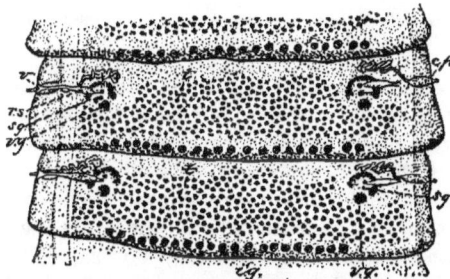

FIG. 118.—Sexually mature segments of the Broad Moniezia (*Moniezia expansa*): *cp*, cirrus pouch; *ig*, interproglottidal glands; *rs*, receptaculum seminis; *sg*, shell gland; *t*, testicles; *v*, vagina; *vg*, vitellogene gland. Enlarged. (After Stiles, 1893, Pl. VI, fig. 4.) See p. 128.

32. Giard's Thysanosoma (*Thysanosoma Giardi*) of Cattle(?), Sheep, and Swine(?).

SYNONYMY.—*Taenia ovilla* Rivolta, 1878 [nec Gmelin, 1790]; *T. Giardi* Moniez, 1879; *T. aculeata* Perroncito, 1882; *Moniezia ovilla* (Rivolta) Moniez, 1891; *M. ovilla* var. *macilenta* Moniez, 1891; *Thysanosoma Giardi* (Moniez) Stiles, 1893; *Th. orilla* (Rivolta) Railliet, 1893; *Taenia Brandti* Cholodkowsky, 1894; *Th. orillum* (Rivolta) Railliet, 1895.

This peculiar tapeworm has been found in sheep in France, Italy, Germany, and Russia, and has been recorded once in hogs; its occurrence as a normal parasite in both hogs and cattle is doubtful. (See pp. 126–127.)

Genus STILESIA.

Two species of this genus are found in sheep, but neither 'form is yet recorded for this continent.

33. TheGlobipunctate Stilesia (*Stilesia globipunctata*) **of Cattle(?) and Sheep.**

SYNONYMY.—*Taenia globipunctata* Rivolta (1874); *T. oripunctata* Rivolta (1874); *Stilesia globipunctata* (Rivolta) Railliet, 1893.

Found in sheep in Italy and India; its occurrence in cattle is doubtful. (See p. 126.)

34. The Centripunctate Stilesia (*Stilesia centripunctata*) **of Cattle(?) and Sheep.**

SYNONYMY.—*Taenia centripunctata* Rivolta (1874); *Stilesia centripunctata* (Rivolta) Railliet, 1893; *Taenia* (*Stilesia*) *centripunctata* of Braun, 1895.

Found in sheep in Italy and Algeria; its presence in cattle is doubtful. (See p. 126.)

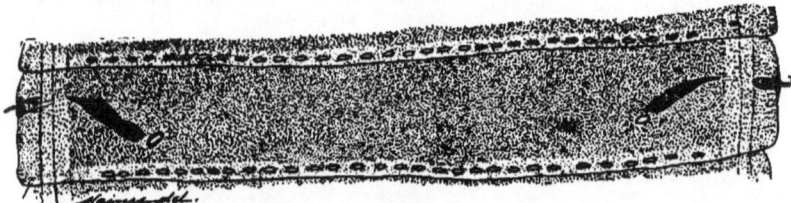

FIG. 119.—Gravid segment of the Broad Moniezia (*Moniezia expansa*), enlarged. (After Stiles, 1893, Pl. VI, fig. 6.) See p. 128.

Life history.—Nothing is positively known about the life history of any of the adult tapeworms of cattle or sheep; but from analogy we may assume that the life cycle is similar to that of other cestodes, namely, that the parasite runs through two stages—the adult form, in the intestine of cattle and sheep, and a larval state (a cysticercus or a cysticercoid), which will be found as a parasite in an intermediate host, probably some invertebrate animal, as an insect, snail, or worm. The intermediate host will become infected from the eggs in the faeces of the cattle and sheep, and the latter will become infected by accidentally swallowing the intermediate host.

While this is what seems to us at present as the probable life history of the bovine and ovine tapeworms, it must be distinctly remembered that no one has as yet been able to positively make out the complete life cycle. In fact, some authors (Curtice and others) do not think that it is necessary for these worms to pass through any intermediate host, but they believe that the embryos (in the eggs) are swallowed by the cattle and develop directly into adult worms. This theory, however, is contrary to analogy, and although this Bureau has repeatedly attempted to infect animals in the manner indicated, none of the experiments can be looked upon as supporting Curtice's views, for we were unable to produce an infection.

One of the following experiments, given as illustrations, might at first sight seem to support Curtice's theory, but can equally well be explained otherwise:

(1) *September 2, 1891.*—A 6-months-old lamb fed with thousands (!!) of eggs of *M. expansa.*

October 2.—Experiment animal showed ripe proglottids in droppings. The infection, however, was totally out of proportion to the number of embryos fed, so that the lamb must have become infected in some other way.

(2) *September 10.*—Lamb fed with thousands of eggs of *M. expansa* at three different times within a week.

September 30.—Lamb killed and four-hour post-mortem held. Intestinal villi, etc., examined microscopically. Result totally negative.

(3) *September 10.*—Lamb fed with thousands of eggs of *M. expansa* five times within a week. Result negative.

Experiments by Curtice and European authors must also be considered as negative, for according to the published accounts of the infections the possible sources besides direct ingestion of eggs were not sufficiently controlled.

TAPEWORM DISEASE OF CATTLE AND SHEEP.

For disease caused by the Fringed Tapeworm, see page 128.

Source of infection.—It will be impossible to make any definite statements upon this point until the complete life history of the worms is known.

Occurrence.—Tapeworms are found in cattle and sheep of all ages and at all times of the year, but calves, lambs, and yearlings suffer more

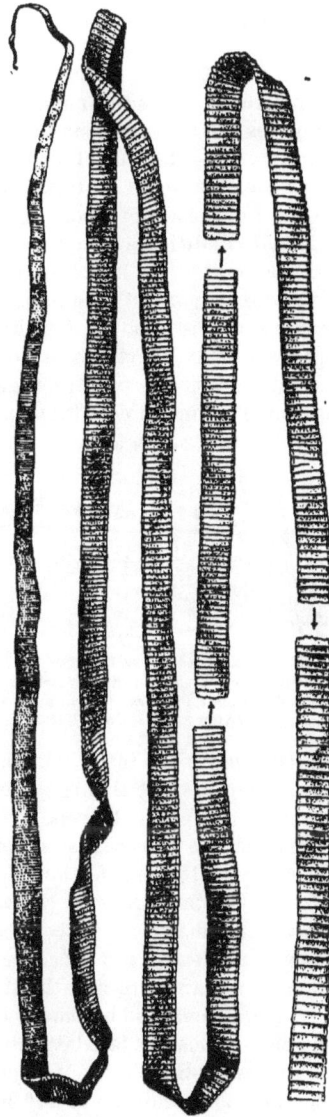

Fig. 120.—Portions of an adult specimen of the Triangle Moniezia (*Moniezia trigonophora*), natural size. (After Stiles & Hassall, 1893, Pl. VIII, fig. 1.) See p. 128.

from the effects of the parasites than do older animals. They are occasionally found in animals in stalls, but are more frequent in animals which are in pasture, and are not so frequent in the winter and early spring as in the summer and fall. Worms (*M. expansa* or *M. planissima*) from 6 to 15 feet long have been found in lambs two to four months old, so that these parasites must grow to maturity very rapidly. Curtice computes the average growth at about 1 yard per month.

Symptoms.—There can be no question that sheep and cattle may harbor a small number of tapeworms with comparatively little or no ill effects, for these worms are found at abattoirs in sheep which are in excellent condition at the time of slaughter. The younger the animal and the greater the infection with worms, the more serious the effects of the disease; but if able to pass through a certain period the animals are very apt to recover, for the worms seem to shed their segments quite suddenly, leaving the hosts with but small tapeworm strobila, and by the time the parasites again attain a greater length the animals may have gained in condition and strength to withstand the disease.

Tapeworms affect their hosts in several ways. By assimilating the

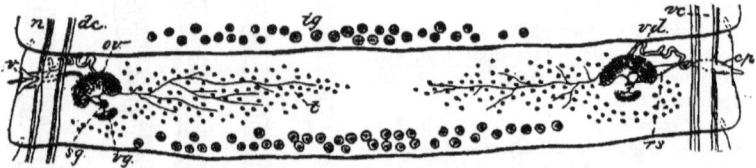

Fig. 121.—Sexually mature segments of the Triangle Moniezia (*Moniezia trigonophora*): *ep*, cirrus pouch; *dc*, dorsal canal; *tg*, interproglottidal glands; *n*, nerve; *ov*, ovary; *rs*, receptaculum seminis; *sg*, shell gland; *t*, testicles; *v*, vagina; *vc*, ventral canal; *vd*, vas deferens; *vg*, vitellogene gland. Enlarged. (After Stiles & Hassall, 1893, Pl. IX, fig. 4.) See p. 128.

nourishment in the intestinal tract of their hosts, they rob the latter of food; when present in large numbers, they may cause stoppage of the bowels, irritate the bowels, leading to non-assimilation of food and reflexly to the nervous symptoms. The clinical history is not very clearly defined from infection with other intestinal parasites, especially with the twisted strongyle (*Strongylus contortus*).

As the animals lose flesh, become poorer, and hidebound, their gait becomes unsteady, the fleece becomes dry and harsh, little yolk being present; the appetite and thirst may increase; diarrhœa is frequent in severe infections, and becomes more pronounced as the disease advances. The animals may at last become completely exhausted and die.

Diagnosis.—Suspicion of tapeworm disease being aroused by the general poor condition of the animals, a positive diagnosis may frequently be made by finding the cast-off segments in the droppings, or around the anus under the tail. A microscopic examination of the faeces for eggs is practicable only for experts. In case of death of one of the flock, it is best to make a careful post-mortem, examining the fourth stomach for the twisted strongyle and the intestines for tapeworms. This can easily be done by opening the intestine in a tub of warm water.

Treatment.[1]—The first thing to do in treating sheep and cattle for tapeworms is to confine the animals in a comparatively small yard and to withhold solid food the night before dosing. The animals should be kept confined until the worms are passed, then all the faeces should be collected and burned, or buried in quicklime.

Schwalenberg reported good results with kamala, dose for a lamb 3.75 grams (about 1 dram); also with cusso (kousso), dose for a lamb 7.5 grams (nearly 2 drams); still better results with kosin (koussin), dose for a lamb 12 centigrams.

Picric acid, dose 0.6 to 1.25 grams (10 to 20 grains), made into pills with meal and water, is recommended by some authors. It should be followed with a cathartic (a 4-ounce dose of Epsom salts or a 4-ounce dose of any of the bland oils).

Two-ounce dose of powdered male fern root, or, still better, the ethereal oil of male fern in dram doses, is recommended by some veterinarians. It can be given in combination with 2 to 4 ounces of castor oil.

Fröhner (1889) gives the following recipes: Take koussin, 3 grains, and of sugar 10 grains, mix, and give at one dose. The dose of tansy is from 2 to 6 drams. It forms one of the chief ingredients of Spinola's worm cake, which is fed to lambs as a preventive against worms. The recipe, sufficient for 100 sheep, is as follows: Take of tansy, calamus root, and tar, each 2½ pounds; of cooking salt, 1¼ pounds; mix these with water and meal, make into cakes, and dry. This is an old and oft-repeated recipe, but I can not vouch for its efficiency. (Curtice, 1890.)

Powdered areca nut may be given to lambs in doses of 1 to 3 drams. If no passage occurs, follow in three or four hours with a cathartic.

In the recent experiments with bluestone by Hutcheon, in South Africa, against wireworm disease in sheep, it has been found that the same treatment expels tapeworms.

Caution.—Repeated accidents have happened from using too strong a solution or too large doses, or in giving it in such a way that the medicine gains access to the lungs. Dr. Hutcheon's method of procedure, which is here given in detail, is safe in the hands of the average farmer if the directions are followed. The person who gives stronger doses than indicated, or who is careless about the measurements, must take the entire responsibility of the miscarriage of the treatment.

FIG. 122.—Adult specimen of the Fringed Tapeworm (*Thysanosoma actinioides*). (After Stiles, 1893, Pl. XI, fig. 1.) See p. 128.

It is a good plan to make up a smaller quantity of the solution and try it upon a few sheep before attempting to dose the entire flock.

---

[1] In this connection consult Curtice, 1890, pp. 120–121.

(*a*) *To prepare the mixture.*—Hutcheon has changed his formula slightly from time to time, the latest published proportions (February 21, 1895) reading as follows (see footnote, p. 136):

Dissolve 1 pound *avoirdupois* (1 pound = 16 ounces) of good, commercial, powdered bluestone (sulphate of copper) in 2 *imperial* quarts (= $2\frac{2}{5}$ quarts U. S.) of boiling water; when the bluestone is *thoroughly* dissolved add $6\frac{1}{2}$ *imperial* gallons (= 26 imperial quarts = $7\frac{4}{5}$ U. S. gallons = $31\frac{1}{5}$ U. S. quarts) of cold water, making in all 7 *imperial* gallons (or $8\frac{2}{5}$ U. S. gallons) of water. (See footnote, p. 136.)

Use only bluestone which is of a uniform blue color; avoid that which is in conglomerate lumps with white patches and covered with a white crust.

The *equivalents* of 1 pound *avoirdupois* and of 7 *imperial* gallons in other weights and measures are as follows:

One pound *avoirdupois* = 1 pound 2 ounces 280 grains of *apothecaries'* or of *imperial troy* weight = 453.59 grams of *metric* weight.

Seven *imperial* gallons = 8 gallons 3 pints 3 fluid ounces 3 fluid drachms 56 minims (or practically 8 gallons $3\frac{1}{2}$ pints, or $8\frac{2}{5}$ gallons) of *apothecaries'* or *wine* measure, U. S. = 31.804409 liters (practically $31\frac{4}{5}$ liters) *metric* system.

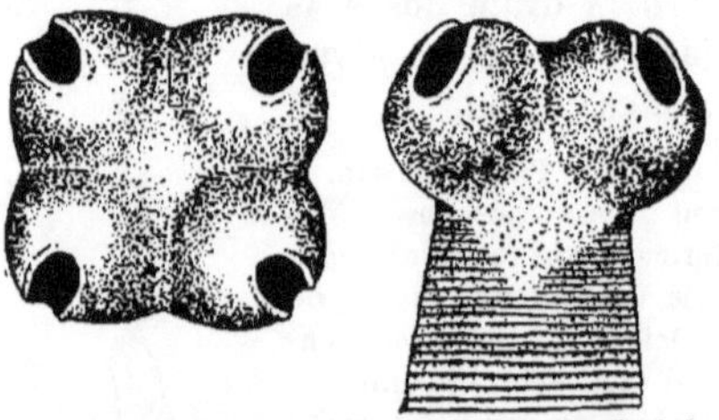

FIG. 123.—Ventral and apex views of the head of the Fringed Tapeworm (*Thysanosoma actinioides*). × 17. (After Stiles, 1893, Pl. XI, figs. 2 and 2 *b*.) See p. 120.

The farmer is cautioned against guessing at the weights and measures, for this is sure to result in too strong a solution, which will kill his animals, or too weak a solution, which will fail to be effective. Scales and measures should be tested before they are used. If reliable scales are not at hand, buy the bluestone already weighed and have the exact weight in avoirdupois, apothecaries', or metric system marked on the package.

If a smaller quantity than the above is desired, this can be made up on the proportion of 1 ounce avoirdupois of bluestone to $4\frac{1}{5}$ U. S. pints of water.

(*b*) *Preparation of the animals.*—Fast the sheep or cattle twenty to twenty-four hours before dosing. If the fast is thirty hours (longer fasts are dangerous) an extra half gallon or a gallon of water should be added to the solution, as animals are more liable to suffer after a long fast.

(*c*) *Size of the dose.*—Hutcheon has several times changed the size of the doses he advises, in some papers basing it on the imperial fluid ounce, in others on the tablespoon. The doses for sheep (in imperial ounces and in tablespoons) given below are his most recent (January 10, 1895) recommendations, and though based upon a solution with 5 per cent less water than the solution given above, they may be used for the weaker mixture.

We have given several of the metric doses to sheep on the Bureau

Experiment Station, and the sheep showed no ill effects; on the contrary they gained in weight. (See footnote, p. 136.)

Age of animals.	Approximate equivalents.			
	Tablespoons. *a*	Imperial.	U. S. apothecaries.	Metric.
For a lamb 3 months old	1	About ¾ fluid ounce.	About ⅔ fluid ounce.	About 20 cc.
For a lamb 6 months old	2	About 1½ fluid ounces.	About 1⅓ fluid ounces.	About 40 cc.
For a sheep 12 months old	3	About 2½ fluid ounces.	About 2 fluid ounces.	About 60 cc.
For a sheep 18 months old	4	About 3 fluid ounces.	About 2⅔ fluid ounces.	About 80 cc.
For a sheep 24 months old	4½	About 3⅓ fluid ounces.	About 3 fluid ounces.	About 90 cc.
For a calf 3 months old	4½ to 5	About 3⅓ to 3¾ fluid ounces.	About 3 to 3⅓ fluid ounces.	90 to 100 cc.
For a calf 6 months old	5 to 5½	About 3¾ to 4⅛ fluid ounces.	About 3⅓ to 3⅔ fluid ounces.	100 to 110 cc.

a "The tablespoon I refer to is the modern full-sized tablespoon (6 fluid drachms). The medicinal tablespoon contains exactly half an ounce."—HUTCHEON.

Be careful not to give a two-toothed young sheep as much as a full grown four-toothed sheep. Mistakes may occur in judging the age unless the teeth are examined.

The doses should be measured off in bottles and the point of each dose plainly marked with a file.

(*d*) *Dosing.*—In dosing, use long-necked bottles, as castor-oil bottles, Worcester sauce bottles, or anchovy sauce bottles.

Let one person set the sheep on its haunches and take its two fore-legs in his left hand, while he steadies the head with the right. Another person inserts the neck of the bottle into the mouth. The head of the sheep should not be raised too high, as in that case the solution may enter the lungs and kill the sheep. A safe rule is to raise the nose to the height of the animal's eyes.

(*e*) *Overdose.*—If, after dosing, any of the sheep seem to be suffering from an overdose, indicated by lying apart from the flock, not feeding, manifesting a painful, excited look and a spasmodic movement in its running, walking with a stiff gait, purging, the discharge being a dirty brownish color, take the affected animals away from the flock to a shady place and dose with laudanum and milk as follows:

For a lamb 4 to 6 months old, 1 teaspoonful of laudanum in a tumbler of milk.

For a sheep 1 year old, 2 teaspoonfuls of laudanum in a tumbler of milk. Repeat half the dose in two to three hours if necessary.

(*f*) *After-treatment.*—The animals should not be allowed water for several hours after receiving their dose.

Prevention.—Preventive measures against adult tapeworm infection in sheep and cattle can be given only in the most general terms, as explicit directions can be based only upon a knowledge of the exact source of infection. The general preventive measures applicable to all intestinal parasitic diseases would apply in the case of tapeworm disease, namely: since the parasites are contracted by means of contaminated food or drink, prevent this contamination as much as possible; feed high with pure food and water preceding and during the time of greatest infection; avoid overcrowding of pastures; isolate infected stock; and when treating medicinally treat the entire flock if possible.

Contamination of food and drink.—This generally takes place by allowing manure piles to drain into the water supply or into pastures In the case of adult tapeworms of cattle and sheep some other factors probably come into play.

Feeding pure food and water.—Grain, etc., should be fed from platforms or troughs, which should be kept clean; raised water troughs should be supplied, so that the animals need not be obliged to drink from stagnant pools. These water troughs should be occasionally cleaned. Many ranchmen have already learned that by feeding their lambs extra grain during the fall, not only have their losses been diminished,

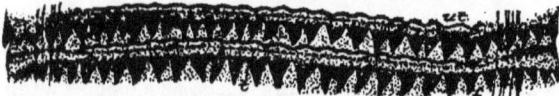

FIG. 124.—Segments of the Fringed Tapeworm (*Thysanosoma actinioides*), showing canals and nerves, and (*f*) fringed border, (*t*) testicles, and (*ut*) uterus. Enlarged. (After Stiles, 1893, Pl. XI, fig. 8.) See p. 128.

but the lambs become larger and stronger as well as fatter.

Avoid overcrowding of pastures.—Overcrowding of pastures is one of the surest methods of keeping animals permanently infested with animal parasites, since the chances of infecting the pasture are increased and, by being compelled to graze too close, the animals are more liable to infection from the germs of parasites found on the ground.

Isolation of infected stock.—This is always advisable, no matter what particular disease is present.

[1] *Treatment of the entire herd.*—This is advisable, since all animals which have been subject to infection stand a chance of having contracted disease, even if only in a light form; but light attacks of parasitic diseases serve to reinfect pastures.

ABATTOIR INSPECTION.

The abattoir inspection for tapeworms in the intestines of cattle and sheep is of no importance whatever, since none of these parasites are transmissible to man in any stage of their development. If the drainage of a slaughterhouse is not properly cared for, the surroundings form a concentrated area of infection.

---

[1] *Addenda to Hutcheon's Bluestone Treatment.*—At the moment of going to press after proof reading was completed, we have received from Hutcheon another article on this subject, dated 1897. He adopts practically the same doses given on p. 135, but changes the strength of the solution (see p. 134) to 1 pound of bluestone to "40 whiskey bottlesful of water." This is practically 1 pound to 7½ *imperial* gallons (=9 Γ. S. gallons = about 34 liters *metric*) of water.

We wish here to repeat and emphasize the advice given to the farmer on p. 133, to make up a smaller quantity of the solution and try it on a few sheep a few days before the entire flock is dosed. This will give him an opportunity to judge whether he has made a mistake in weights and measures in mixing the solution.

II. COMPENDIUM OF THE PARASITES, ARRANGED ACCORDING TO THEIR HOSTS.

By Albert Hassall.

In the following compendium are included the hosts for all of the parasites discussed in this paper. The numbers of the hosts refer to the numbers in von Linstow's (1878) Compendium. In selecting the scientific names of hosts, I have been guided by the advice of Dr. T. S. Palmer, of the Biological Survey, U. S. Department of Agriculture.

⊞ signifies that either Stiles or I have examined the parasite for the host in question in North America.

◻ signifies that either Stiles or I have examined this parasite for the host in question, but the specimen was not North American.

? signifies that I doubt the validity of the determination or the validity of the species.

† signifies that I reserve judgment upon the species.

MAMMALS (*Mammalia*).

PRIMATES.

1. **Homo sapiens. Man.**
     Dicrocoelium lanceatum, p. 55 .... Liver.
     Fasciola hepatica, p. 29 .... Liver.
     † *Fasciola hepatica angusta*, p. 48 .... Lungs.
     † *Fasciola gigantica*, p. 49 .... Lungs.
     ⊞ *Schistosoma haematobium*, p. 58 .... Veins.
     Bothriocephalus cordatus, p. 85 .... Intestine.
     ⊞ *Bothriocephalus latus*, p. 85 .... Intestine.
     Bothriocephalus Mansoni, p. 85 .... Intestine.
     ⊞ *Cysticercus cellulosae*, p. 89 .... Muscles, eye, and brain.
     † *Cysticercus tenuicollis*, p. 96 .... Omentum.
     Davainea madagascariensis, p. 86 .... Intestine.
     Dipylidium caninum, p. 86 .... Intestine.
     ⊞ *Echinococcus polymorphus*, p. 113 .... Especially liver and lungs.
     ⊞ *Hymenolepis diminuta* (including *Taenia flavopunctata*), p. 86 ...... Intestine.
     ◻ *Hymenolepis murina* (including *Taenia nana*), p. 86 .... Intestine.
     Krabbea grandis .... Intestine.
     ⊞† *Taenia confusa*, p. 85 .... Intestine.
     ⊞ *Taenia saginata*, p. 71 .... Intestine.
     ⊞ *Taenia solium*, p. 89 .... Intestine.
   Simia faunus.
     Cysticercus tenuicollis, p. 96.
14. **Simia inuus.** (*See* **Macacus inuus.**)
9. **Simia rubra.** (*See* **Cercopithecus patas.**)

137

19. Simia silenus. (*See* **Macacus silenus.**)
   Semnopithecus entellus. Hanuman langur.
     Cysticercus tenuicollis, p. 96.
4. Cercopithecus cephus.
     Cysticercus cellulosae, p. 89..................................... Peritoneum.
5. Cercopithecus cynosurus. Malbrouck Guenon.
     Cysticercus tenuicollis, p. 96......................... Liver and mesentery.
6. Cercopithecus fuliginosus. Sooty Monkey.
     Schistosoma haematobium (Cobbold's *Bilharzia magna*), p. 58......... Veins.
7. Cercopithecus mona. Mona Guenon.
     Cysticercus tenuicollis, p. 96.
9. Cercopithecus patas. Patas Guenon.
     Cysticercus cellulosae, p. 89.
10. Cercopithecus sabaeus. Grivet Guenon.
     Cysticercus tenuicollis, p. 96.......................... Liver and mesentery.
12. Macacus cynomolgus. Crab-eating Macaque.
     Cysticercus tenuicollis, p. 96......................... Liver and mesentery.
     Echinococcus polymorphus, p. 113.................................. Viscera.
14. Macacus inuus. Barbary Macaque.
     Cysticercus cellulosae, p. 89................................... Peritoneum.
     Cysticercus tenuicollis, p. 96.................................. Peritoneum.
     Echinococcus polymorphus, p. 113.................................. Viscera.
19. Macacus silenus. Lion-tailed Macaque.
     Echinococcus polymorphus, p. 113.................................. Viscera.
12. Inuus cynomolgus. (*See* **Macacus cynomolgus.**)
14. Inuus ecaudatus. (*See* **Macacus inuus.**)
17. Papio maimon. Mandril.
     Cysticercus tenuicollis, p. 96.......................... Liver and mesentery.
17. Cynocephalus maimon. (*See* **Papio maimon.**)

CARNIVORES (*Carnivora*).

191. Ursus arctos. Brown Bear.
     Cysticercus cellulosae, p. 89..................................... Muscles.
167. Vulpes lagopus. Arctic Fox.
     Taenia coenurus, p. 109.. Intestine.
165. Canis familiaris. Dog.
     Bothriocephalus cordatus, p. 101................................. Intestine.
     Bothriocephalus fuscus, p. 101................................... Intestine.
     Bothriocephalus latus, p. 101.................................... Intestine.
     Bothriocephalus serratus, p. 101................................. Intestine.
     Cysticercus cellulosae, p. 89.................... Muscles and peritoneum.
     ⊞ *Dipylidium caninum,* p. 102.................................... Intestine.
     ☐ *Mesocestoides lineatus,* p. 102................................ Intestine.
     Taenia coenurus, p. 109.. Intestine.
     ⊞ *Taenia echinococcus,* p. 114................................... Intestine.
     ☐ *Taenia Krabbei,* p. 102... Intestine.
     ⊞ *Taenia marginata,* p. 96....................................... Intestine.
     ⊞ *Taenia serialis,* p. 102.. Intestine.
     ⊞ *Taenia serrata,* p. 102... Intestine.
146. Felis domestica. Cat.
     ⸮ *Dicrocoelium lanceatum,* p. 55........................... Gall bladder.
     Fasciola hepatica, p. 29.
     Cysticercus cellulosae, p. 89.
     Cysticercus tenuicollis, p. 96.
     Echinococcus polymorphus, p. 113.

RODENTS (*Rodentia*).

Lepus californicus.
 Coenurus serialis, p. 102.
Lepus callotis.
 ⊞ *Coenurus serialis*, p. 102.
137. **Lepus cuniculus. European Wild Rabbit.**
 Dicrocoelium lanceatum, p. 55.
 Fasciola hepatica, p. 29.
 □ *Coenurus serialis*, p. 102.
 ? *Coenurus cerebralis*, p. 109.
 Echinococcus polymorphus, p. 113.
137*a*. **Lepus cuniculus domesticus. Common domesticated Rabbit.**
 Fasciola hepatica, p. 29... Liver.
 ? *Coenurus cerebralis*, p. 109.. Muscles.
 Coenurus serialis, p. 102... Muscles.
140. **Lepus timidus. Common European Hare.**
 Dicrocoelium lanceatum, p. 55.............................. Gall bladder.
 Fasciola hepatica, p. 29... Liver.
 ? *Coenurus cerebralis*, p. 109... Muscles.
 Coenurus serialis, p. 102... Muscles.
139. **Lepus variabilis. Mountain Hare.**
 Dicrocoelium lanceatum, p. 55.. Liver.
 Coenurus serialis, p. 102.................................... Connective tissue.
 Cavia cobaya. Guinea Pig.
 Fasciola hepatica caviae, p. 48.. Liver.
110. **Mus rattus. Black Rat.**
 Cysticercus cellulosae, p. 89.. Peritoneum.
98. **Castor fiber. European Beaver.**
 Fasciola hepatica, p. 29... Liver.
87. **Sciurus cinereus.**
 Cysticercus tenuicollis, p. 96.................... Liver and mesentery.
86. **Sciurus vulgaris. European Squirrel.**
 Fasciola hepatica, p. 29... Liver.
 Cysticercus tenuicollis, p. 96...................... Liver and mesentery.

UNGULATES (*Ungulata*).

206. **Elephas indicus. Indian Elephant.**
 Fasciola hepatica, p. 29... Liver.
248. **Equus caballus. Horse.**
 □ *Fasciola hepatica*, p. 29.. Liver.
 Coenurus cerebralis, p. 109.. Cerebrum.
 ⊔ *Echinococcus polymorphus*, p. 113.................................... Liver.
246. **Equus asinus. Ass.**
 Dicrocoelium lanceatum, p. 55.. Liver.
 □ *Fasciola hepatica*, p. 29.. Liver.
 □ *Echinococcus polymorphus*, p. 113..................................... Liver.
215, **Bos bubalis. Indian Buffalo.**
 Amphistoma cervi, p. 64 .... Rumen.
 □ *Fasciola hepatica aegyptiaca*, p. 48................................... Liver.
 □ *Gastrothylax gregarius*, p. 67.. Rumen.
217*b*. **Bos frontalis. Gayal.**
 Gastrothylax Cobboldii, p. 67.. Rumen.
 Gastrothylax elongatum, p. 67... Rumen.
 Homalogaster Paloniae, p. 67.. Caecum.
 Bos indicus. Zebu. (See also p. 67).
 Moniezia expansa, p. 128.. Intestine.

216. Bos taurus. Domesticated cattle.

☐ *Amphistoma cervi*, p. 64.. Rumen.
Amphistoma explanatum, p. 67.............................. Gall bladder.
Amphistoma tuberculatum, p. 67................................... Rumen.
☐ *Dicrocoelium lanceatum*, p. 55................................... Liver.
Dicrocoelium pancreaticum, p. 57 .... Pancreas.
❢ *Fasciola gigantica*, p. 49.. Liver.
⊞ *Fasciola hepatica*, p. 29.............................. Liver and lungs.
☐ *Fasciola hepatica angusta*, p. 48............................ Liver.
☐ *Fasciola hepatica aegyptiaca*, p. 48...................... Liver.
⊞ *Fasciola magna*, p. 49.............................. Liver and lungs.
Gastrothylax crumenifer, p. 67......................... Rumen.
Homalogaster Poirieri, p. 67........................ Large intestine.
☐ *Schistosoma bovis*, p. 60.......................... Veins.
❢ *Schistosoma haematobium*, p. 58...................... Veins.
Coenurus cerebralis, p. 109............................ Brain.
⊞ *Cysticercus bovis*, p. 71............................ Muscle.
⊞ *Cysticercus tenuicollis*, p. 96.............. Liver and mesentery.
⊞ *Echinococcus polymorphus*, p. 113.............. Liver and lungs.
☐ *Moniezia alba*, p. 127............................ Intestine.
☐ *Moniezia Benedeni*, p. 128........................ Intestine.
⊞ *Moniezia expansa*, p. 128......................... Intestine.
⊞ *Moniezia planissima*, p. 127...................... Intestine.
❢ *Stilesia centripunctata*, p. 130................. Intestine.
❢ *Stilesia globipunctata*, p. 130.................. Intestine.
❢ *Thysanosoma Giardi*, p. 129...................... Intestine.

218. Ovibos moschatus. Musk Ox.

Moniezia expansa, p. 128......................... Intestine.

Ovis ammon.

Echinococcus polymorphus, p. 113.

219. Ovis argali. Argali.

Fasciola hepatica, p. 29.......................... Liver.
Cysticercus tenuicollis, p. 96.......... Liver and mesentery.
Echinococcus polymorphus, p. 113.......... Liver and lungs.

220. Ovis aries. Domesticated sheep.

⊞ *Amphistoma cervi*, p. 64............................ Rumen.
☐ *Dicrocoelium lanceatum*, p. 55...................... Liver.
⊞ *Fasciola hepatica*, p. 29.......................... Liver.
Dicrocoelium pancreaticum, p. 57.
Schistosoma bovis, p. 60.......................... Veins.
☐ *Coenurus cerebralis*, p. 109............. Brain and spinal cord.
⊞ *Cysticercus tenuicollis*, p. 96................. Mesentery.
☐ *Echinococcus polymorphus*, p. 113.......... Liver and lungs.
Moniezia alba, p. 127............................ Intestine.
☐ *Moniezia Benedeni*, p. 128...................... Intestine.
⊞ *Moniezia expansa*, p. 128....................... Intestine.
☐ *Moniezia Neumanni*, p. 128 .... Intestine.
❗☐ *Moniezia nullicollis*, p. 26.................. Intestine.
⊞ *Moniezia planissima*, p. 127................... Intestine.
⊞ *Moniezia trigonophora*, p. 128................. Intestine.
☐ *Moniezia Vogti*, p. 127......................... Intestine.
☐ *Stilesia centripunctata*, p. 130 .... Intestine.
☐ *Stilesia globipunctata*, p. 130................. Intestine.
⊞ *Thysanosoma actinioides*, p. 128........ Gall ducts and intestine.
☐ *Thysanosoma Giardi*, p. 129.................... Intestine·

220. **Ovis laticauda.**

    ☐ *Moniezia expansa*, p. 128.. Intestine

221. **Ovis musimon. Mufflon.**

    Coenurus cerebralis, p. 109 .... Cerebrum.

    Cysticercus tenuicollis, p. 96.......................... Liver and mesentery.

222. **Capra hircus. Goat.**

    Amphistoma cervi, p. 64 .... Rumen.

    Dicrocoelium lanceatum, p. 55.. Liver.

    Fasciola hepatica, p. 29.. Liver.

    Coenurus cerebralis, p. 109.. Brain.

    ⊞ *Cysticercus tenuicollis*, p. 96 .... Mesentery.

    Echinococcus polymorphus, p. 113.................. Liver and lungs.

    Moniezia expansa, p. 128... Intestine.

    ! ☐ *Moniezia caprae*.. Intestine.

    Capra pyrenaica. Spanish Ibex.

    Moniezia expansa, p. 128... Intestine.

225. **Rupicapra tragus. Chamois or Gemse.**

    Cysticercus tenuicollis, p. 96..................... Liver and mesentery.

    Moniezia expansa, p. 128... Intestine.

    Boselaphus tragocamelus. Nilgai or Blue Bull.

    Fasciola hepatica, p. 29... Liver.

    Fasciola magna, p. 49... Liver.

    Hippotragus equinus. Roan Antelope.

    Coenurus cerebralis, p. 109.

    Oryx beisa. Beisa.

    Cysticercus tenuicollis, p. 96.

224. **Oryx leucoryx. Leucoryx.**

    Cysticercus tenuicollis, p. 96....................... Liver and mesentery.

226. **Saiga tartarica. Saiga.**

    Cysticercus tenuicollis, p. 96....................... Liver and mesentery.

223. **Gazella dorcas. Dorcas Gazelle.**

    Amphistoma cervi, p. 64 .... Rumen.

    Dicrocoelium lanceatum, p. 55.. Liver.

    Fasciola hepatica, p. 29.. Liver.

    Cysticercus tenuicollis, p. 96.................................... Mesentery.

    Moniezia expansa, p. 128... Intestine.

227. **Gazella euchore. Springbok.**

    Cysticercus tenuicollis, p. 96..................... Liver and mesentery.

227. **Antilope euchore.** (*See* **Gazella euchore.**)

224. **Antilope leucoryx.** (*See* **Oryx leucoryx.**)

226. **Antilope saiga.** (*See* **Saiga tartarica.**)

    Kobus ellipsiprymus. Waterbuck.

    Cysticercus tenuicollis, p. 96.

    Antilocapra americana.

    ⊞ *Cysticercus bovis*, p. 71.. Muscles.

241. **Giraffa camelopardalis. Giraffe.**

    Fasciola gigantica, p. 49 .... Liver.

    Cysticercus bovis, p. 71.. Muscles.

    Echinococcus polymorphus, p. 113 .... Liver.

240. **Cariacus americanus. Virginia Deer.**

    Fasciola hepatica, p. 29... Liver.

    ⊞ *Fasciola magna*, p. 49.. Liver.

232. **Cariacus campestris.**

    Amphistoma cervi, p. 64.. Rumen.

    Moniezia expansa, p. 128.. Intestine.

237. Cariacus nambi.
Amphistoma cervi, p. 64.. Rumen.
Moniezia expansa, p. 128.. Intestine.
Thysanosoma actinioides, p. 128.. Intestine.
233. Cariacus paludosus.
Amphistoma cervi, p. 64.. Rumen.
Thysanosoma actinioides, p. 128.. Intestine.
238. Cariacus rufus. Brocket.
Amphistoma cervi, p. 64.. Rumen.
Cysticercus tenuicollis, p. 96............................ Liver and mesentery.
Moniezia expansa, p. 128.. Intestine.
Thysanosoma actinioides, p. 128.. Intestine.
238. Mazama rufus. (*See* **Cariacus rufus.**)
Cariacus simplicicornis.
Amphistoma cervi, p. 64... Rumen.
Cysticercus tenuicollis, p. 96............................ Liver and mesentery.
Thysanosoma actinioides, p. 128.. Intestine.
240. Cariacus virginianus. (See **Cariacus americanus.**)
234. Capreolus caprea. Roe Deer.
Amphistoma cervi, p. 64.. Rumen.
Fasciola hepatica, p. 29.. Liver.
Coenurus cerebralis, p. 109.. Brain.
Cysticercus cellulosae, p. 89............................ Liver and mesentery.
Cysticercus tenuicollis, p. 96.
Moniezia expansa, p. 128.. Intestine.
Taenia crucigera.. Intestine.
230. Alce alces. European Elk.
Amphistoma cervi, p. 64.. Rumen.
Echinococcus polymorphus, p. 113.
230. Alces machlis. (*See* **Alce alces.**)
230. Alces palmatus. (*See* **Alce alces.**)
239. Tarandus rangifer. Reindeer or Caribou.
Coenurus cerebralis, p. 109.. Brain.
Cysticercus tenuicollis, p. 96............................ Liver and mesentery.
230. Cervus alces. (*See* **Alce alces.**)
231. Cervus axis. Axis deer.
Cysticercus tenuicollis, p. 96............................ Liver and mesentery.
Cervus canadensis. Elk or Wapiti.
Fasciola magna, p. 49.. Liver.
235. Cervus dama. Fallow Deer.
Amphistoma cervi, p. 64.. Rumen.
Dicrocoelium lanceatum, p. 55.. Liver.
Fasciola hepatica, p. 29.. Liver.
□ *Fasciola magna*, p. 49.. Liver.
236. Cervus elaphus. Stag.
Amphistoma cervi, p. 64.. Rumen.
Dicrocoelium lanceatum, p. 55.. Liver.
Fasciola hepatica, p. 29.. Liver.
Fasciola magna, p. 49.. Liver.
Cysticercus tenuicollis, p. 96............................ Liver and mesentery.
Cervus tarandus. (*See* **Tarandus rangifer.**)
Cervus unicolor. Sanbur, Rusa Deer.
Coenurus cerebralis, p. 109.. Brain.
Cysticercus tenuicollis, p. 96............................ Liver and mesentery.

244. **Auchenia llama. Llama.**
Dicrocoelium lanceatum, p. 55.. Liver.
⊞ *Cysticercus bovis*, p. 71 .... Muscles.
243. **Camelus bactrianus. Bactrian Camel.**
Fasciola hepatica, p. 29... Liver.
⊞ *Echinococcus polymorphus*, p. 113......................... Liver and viscera.
243. **Camelus dromedarius. Dromedary.**
Coenurus cerebralis, p. 109.. Brain.
Echinococcus polymorphus, p. 113................................ Viscera.
213. **Phachochoerus africanus. Aelian's Wart Hog.**
Cysticercus tenuicollis, p. 96.
214. **Phachochoerus aethiopicus. Pallas' Wart Hog.**
Cysticercus tenuicollis, p. 96.................................... Abdomen.
210. **Potamochoerus porcus. Red River Hog.**
Cysticercus tenuicollis, p. 96 .... Omentum.
210. **Potamochoerus penicillatus.** (*See* **Potamochoerus porcus.**)
208. **Sus scrofa. Wild boar.**
Cysticercus cellulosae, p. 89................................... Muscles.
Cysticercus tenuicollis, p. 96 .... Omentum.
209. **Sus scrofa domestica. Domesticated swine.**
Agamodistomum suis, p. 28 .... Muscles.
Dicrocoelium lanceatum, p. 55.............................. Liver.
Fasciola hepatica, p. 29..................................... Liver.
⊞ *Cysticercus cellulosae*, p. 89............................ Muscles.
⊞ *Cysticercus tenuicollis*, p. 96 .... Omentum.
Thysanosoma Giardi, p. 129................................ Intestine.

CETACEANS (*Cetacea*).

Orca gladiator. Grampus or Killer.
☐ *Fasciola hepatica*, p. 29 .... Liver.

MARSUPIALS (*Marsupialia*).

283. **Macropus giganteus. Gray Kangaroo.**
Fasciola hepatica, p. 29.. Liver.
Echinococcus polymorphus, p. 113.
Macropus major.
Echinococcus polymorphus, p. 113.

MOLLUSKS (*Mollusca*).

Limnaea oahuensis.
Fasciola hepatica, p. 29.
Limnaea rubella.
Fasciola hepatica, p. 29.
1872. **Limnaea truncatula.**
Fasciola hepatica, p. 29.
1874b. **Physa alexandrina.**
Cercaria pigmentata (*Amphistoma cervi*), p. 64.
1874c. **Physa micropleura.**
Cercaria pigmentata (*Amphistoma cervi*), p. 64.

III. BIBLIOGRAPHY OF THE MORE IMPORTANT WORKS CITED.

By ALBERT HASSALL.

BASSI, R.
    1875.—Sulla cachessia itteroverminosa o marciaia dei cervi, causata dal *Distomum magnum* < Il Medico Veterinario, pp. 497–513, Tav. I–III, Torino.

BATSCH, A. J. G. C.
    1786.—Naturgeschichte der Bandwurmgattung überhaupt und ihrer Arten insbesondere, nach den neuern Beobachtungen in einem systematischen Auszuge. 298 pp., Tabs. I–V. Halle.

BILLINGS, J. S.
    1885.—Index-Catalogue of the Library of the Surgeon-General's Office, United States Army. Vol. VI, 11 + 1051 pp. Washington.

BITTING, A. W.
    1895.—Liver Fluke. Leeches in the Liver < Bulletin No. 28, Florida Agricultural Experiment Station, pp. 83–85, Pls. I–II, 1 map.

BLANCHARD, R.
    1885.—Traité de zoologie médicale. Tom. I, fasc. I, pp. 1–192. Paris.
    1886.—Traité de zoologie médicale. Tom. I, fasc. II, pp. 193–480. Paris.
    1888.—Traité de zoologie médicale. Tom. I, fasc. III, pp. 481–808. Paris.
    1895.—Les Hématozoaires de l'homme et des animaux, 208 pp., 11 figs. Paris.
    1895.—Maladies parasitaires. Parasites animaux, parasites végétaux. A l'exclusion des bactéries < Traité de pathologie générale (Bouchard), Tom. II, pp. 649–932, figs. 47–116.

BRAUN, MAX.
    1889.—Vermes < Bronn's Klassen und Ordnungen des Thier-Reichs. Bd. IV, Lief. 9–11, pp. 305–400, Taf. VI–VIII.
    1895.—Die thierischen Parasiten des Menschen. 283 pp., 147 figs. Würzburg.

CHOLODKOWSKY, N.
    1894.—Ueber eine neue species von Taenia. < Cent. f. Bakt. u. Paras., XV, pp. 552–554, 2 figs.

COBBOLD, T. SPENCER.
    1864.—Entozoa: An introduction to the study of Helminthology, with reference, more particularly, to the internal parasites of man. 480 pp. XXI plates, 82 figs. in text. London.
    1875.—Further remarks on Parasites from the Horse and Elephant, with a notice of new Amphistomes from the Ox < The Veterinarian, Vol. XLVIII, pp. 817–821.

CREPLIN, F. C. H.
    1837.—Art. Distoma < Ersch und Gruber's Allg. Encycl., 1 sect., 29. Th., pp. 309–329.
    1847.—Beschreibung zweier neuen Amphistomen-Arten aus dem Zebuochsen < Arch. f. Naturg., XIII. Jhg., I. Bd., pp. 30–35, Taf. II, figs. 1–5.

CURTICE, C.
    1890.—The Animal Parasites of Sheep. U. S. Dept. Agric., Bureau of Animal Industry. 222 pp., 36 plates. Washington.

DAVAINE, C.
    1877.—Traité des Entozoaires et des maladies vermineuses de l'homme et des animaux domestiques. 2. édit. Pp. cxxxii, 72 figs., pp. 1003, 38 figs. Paris.

DEFFKE, O.
    1891.—Die Entozoen des Hundes < Arch. f. wiss. u. prakt. Thierheilk. Bd.
        XVII, pp. 1-60, 253-289, Taf. I-II.
DEWITZ, J.
    1892.—Die Eingeweidewurmer der Haussäugethiere. 180 pp., 141 figs. Berlin.
DIESING, K. M.
    1850.—Systema Helminthum, I. 680 pp. Vindobonnae.
    1858.—Revision der Myzhelminthen. Abtheilung: Trematoden < Sitz. d. math.-
        nat. Cl. d. k. Akad. d. Wiss., Wien, Bd. XXXII, pp. 307-390, Taf. I-II.
DINWIDDIE, R. R.
    1889.—Veterinarian's Report. Second Annual Report of the Arkansas Agricul-
        tural Experiment Station. Pp. 109-119, 1 fig.
    1892.—Some Parasitic Affections of Cattle < Journ. Comp. Med. and Vet. Arch.,
        XIII (6), June, pp. 342-343.
DUJARDIN, F.
    1845.—Histoire naturelle des Helminthes, ou vers intestinaux. 654 pp., Pls. I-XII.
        Paris.
DUNCKER, H. C. F.
    1896.—Die Muskeldistomeen < Berl. thieriirztl. Wochenschr., No. 24, pp. 279-
        282, 6 figs.
FRANCIS, M.
    1891.—Liver Flukes < Texas Agric. Exp. Sta. Bulletin No. 18, 9 pp., 6 figs.
FRENCH, C.
    1896.—On Intestinal Parasitism in the Dog, and its Treatment < The Journ. Comp.
        Med. and Vet. Arch., Vol. XVII (6), June, pp. 441-452.
FRIIS, ST.
    1897.—Om Kødkontrollens Standpunkt over for tintet Oksekød < Maanedsskrift
        for Dyrlaeger, Bd. IX, Heft 2-3, pp. 83-87.
GIARD, A., & BILLET, A.
    1892.—Sur quelques Trématodes des bœufs du Tonkin < C. R. Soc. Biol., 9. s-r.,
        IV (25), 8 juillet, pp. 613-615.
GLAGE.
    1896.—Versuche über Tötung von Finnen durch elektrische Ströme < Zeit. f.
        Fleisch-u. Milchhyg., VII Jhg., Heft 2, pp. 21-26.
    1896.—Versuche über die Lebenszähigkeit der Finnen < Zeit. f. Fleisch-u. Mil-
        chhyg., VI Jhg., Heft 2, pp. 231-234.
GMELIN, J. F.
    1790.—Systema Naturae. Tom. I, Pars VI.
GOEZE, J. A. E.
    1782.—Versuch einer Naturgeschichte der Eingeweidewürmer thierischer Körper
        471 pp., 35 Taf. Blankenburg.
GOUVEA, H.
    (1895).—La distomatose pulmonaire par la douve du foie. Thèse de Paris,
        No. 104.
HARLEY, J.
    1864.—On the Endemic Haematuria of the Cape of Good Hope < Med.-Chir. Trans.
        London, 2. ser., XLVII, pp. 55-72, Pl. II, IIa.
HASSALL, A.
    1891.—A New Species of Trematode infesting Cattle (F. carnosa) < American
        Vet. Rev., XV, July, pp. 208-209, 1 fig.
    1891.—Fasciola americana < American Vet. Rev., XV, September, p. 359.
    1894.—(See Stiles 1894-95.)
HUBER, J. Ch.
    1890.—Zur Litteraturgeschichte der Leberegelkrankheit < Deutsche Zt. f.
        Thiermed. u. vergl. Path., XVII, pp. 77-79.

HUBER, J. Ch.—Continued.

1891.—Echinococcus cysticus <Bibliographie der klinischen Helminthologie, Heft 1, pp. 1-39, München.

1892.—Die Darmcestoden des Menschen <Bibliographie der klinischen Helminthologie, Heft 3, No. 4, pp. 69-150, München.

1894.—Trematoden <Bibliographie der klinischen Helminthologie, Heft 7-8, pp. 283-287.

JANSON, J. L.

1893.—Die Hausthiere in Japan. IV. Die Krankheiten der Hausthiere in Japan <Arch. f. wiss. u. prakt. Thierheilkunde, XIX, pp. 241-276.

KRABBE, H.

1865.—Helminthologiske Undersøgelser i Danmark og paa Island, med saerligt Hensyn til Blaereormlidelserne paa Island. 64 pp., 7 plates. Kjøbenhavn.

LEIDY, J.

1856.—A Synopsis of Entozoa and some of their Ecto-congeners observed by the Author <Proc. Acad. Nat. Sci. Phila., VIII, pp. 42-58.

1891.—Notes on Entozoa <Proc. Acad. Nat. Sci. Phila., pp. 234-236.

LEUCKART, R.

1863.—Die menschlichen Parasiten und die von ihnen herrührenden Krankheiten. I. Bd., viii+766 pp., 268 figs. Leipzig und Heidelberg.

1879.—Die Parasiten des Menschen und die von ihnen herrührenden Krankheiten. 2. Aufl., I. Bd., 1. Lief., pp. i-viii+1-336.

1880 —Die Parasiten des Menschen und die von ihnen herrührenden Krankheiten. 2. Aufl., I. Bd., 2. Lief., pp. i-xii+337-856.

1881-1882.—Zur Entwickelungsgeschichte des Leberegels <Zool. Anz., IV, pp. 641-646; 1882, V, pp. 524-528.

1886.—Die Parasiten des Menschen und die von ihnen herrührenden Krankheiten. 2. Aufl., I. Bd., 3. Lief., pp. i-xxxi+855-1000; 2. Aufl., I. Bd., 2. Abth., pp. 1-96, i-iv.

1889.—Die Parasiten des Menschen und die von ihnen herrührenden Krankheiten. 2. Aufl., I. Bd., 2. Abth., 4. Lief., pp. 97-440.

1892.—Ueber den grossen amerikanischen Leberegel <Centralbl. für Bakt. u. Paras., XI (25), 16. Juni, pp. 797-799.

1894.—Die Parasiten des Menschen und die von ihnen herrührenden Krankheiten. 2. Aufl., I. Bd., 2. Abth , 5. Lief., pp. i-viii+441-736.

LINNAEUS, C.

1758.—Caroli Linnaei Systema naturae regnum animale. 10. ed.

LINSTOW, O. von.

1878.—Compendium der Helminthologie. Hannover.

1880.—Compendium der Helminthologie. Nachtrag. Hannover.

LOOSS, A.

1895.—Zur Anatomie und Histologie der *Bilharzia haematobia* (Cobbold) <Arch. f. mikroskop. Anat., Bd. XLVI, pp. 1-108, Taf. I-III.

1896.—Recherches sur la faune parasitaire de l'Égypte. Première partie <Mém. de l'Institut Égyptien, III, pp. 1-252, Pls. I-XVI, Cairo.

LUNGWITZ, J. M.

1895.—*Taenia orilla* Rivolta, Anatomischer Bau und die Entwickelung ihrer Geschlechtsorgane <Arch. f. wiss. u. prakt. Thierheilk., XXI (2-3), pp. 105-159, Taf. II-III.

LUTZ, A.

1892.—Zur Lebensgeschichte des *Distoma hepaticum* <Centralbl. f. Bakt. u. Paras., XI (25), 16. Juni, pp. 783-796.

1893.—Weiteres zur Lebensgeschichte des *Distoma hepaticum* <Centralbl. f. Bakt. u. Paras., XIII (10), 13. März, pp. 320-328.

Meiilis, E.
1825.—Observationes anatomicae de distomate hepatico et lanceolato. Gotting. fol. 1 Tab.
Naunyn, B.
1863.—Ueber die zu *Echinococcus hominis* gehörige Tänie <Arch. f. Anat., Phys. u. wiss. Med., pp. 412-416, Taf. x, B, figs. 1-4.
Neisser, A.
1877.—Die Echinococcen-Krankheit. 228 pp. Berlin.
Neumann, L. G.
1892 A.—Traité des maladies parasitaires non microbiennes des animaux domestiques. 767 pp., 364 figs. Paris.
1892 B.—A Treatise on the Parasites and Parasitic Diseases of the Domesticated Animals. Translated and edited by George Fleming from 2d French edition. 800 pp., 364 figs. London.
Ostertag, R.
1895.—Handbuch der Fleischbeschau für Tierärzte, Ärzte und Richter. 2. Aufl. xvi+733 pp., 161 figs. Stuttgart.
1896.—Ueber das Vorkommen der Rinderfinnen und die Verwertung des Fleisches der finnigen Rinder in den Grössern Norddeutschen Schlachthöfen < Zeit. f. Fleisch-u. Milchhyg., VI, Jhg., Heft 6, 8, 12. pp. 103-107, 143-149, 227-230.
1897.—Beitrag zur Frage der Entwickelung der Rinderfinnen und der Selbst-Heilung der Rinderfinnenkrankheit <Zeit. f. Fleisch-u. Milchhyg, VIII, Jhg., Heft 1 (Oct.), pp. 1-4.
Otto, R.
1896.—Beiträge zur Anatomie und Histologie der Amphistomeen <Deuts. Zeit. f. Thiermed., XXII, pp. 85-141, figs. 1-17; 275-296, figs. 18-30.
Poirier, J.
(1883).—Description des Helminthes nouv. du *Palonia frontalis* <Bullet. Soc. philomat., 7. sér., VII, pp. 73-80, Pl. ii.
Raillet, A.
1893.—Traité de Zoologie médicale et agricole. Fasc. 1, pp. 1-736, figs. 1-497.
1895.—Sur une forme particulière de douve hépatique provenant du Sénégal <C. R. Soc. Biol., 10. sér., II (15), 10 mai, pp. 338-340.
1896.—Sur quelques parasites du dromadaire <C. R. Soc. Biol., 10. sér., III (17), 22 mai, pp. 489-492.
1897.—La Douve pancréatique <Rec. d. Méd. Vét., 8 sér., T. IV, No. 14, pp. 371-377, 1 fig.
Rassmussen, P. B.
1897.—Om okse-og svinetinten <Norsk Veterinaer-Tidsskrift., IX, ii og iii, pp. 33-77; also <Maanedsskrift for Dyrlaeger, IX, ii-iii, pp. 33-83.
Reissman.
1897.—Referat [of Vollers, Noack, Zschokke, Foth, Glage] <Hygieuische Rundschau, VII (19), 1 Oct., pp. 966-973.
In this Review, which reached us after our proof reading, Reissman adds some interesting observations of his own. He maintains that four to five days at a temperature of —7° C. to —8° C. is ample to insure the death of pork measles. The loss of weight in hams in freezing is slightly less than 2 per cent.
Rudolphi, K. A.
1803.—Neue Beobachtungen über die Eingeweidewürmer <Arch. f. Zool. und Zoot., III, ii, pp. 1-32.
1809.—Entozoorum sive vermium intestinalium historia naturalis. Amstelaedami, II, ii.
1819.—Entozoorum synopsis cui accedunt Mantissa duplex et Indices. Vindobonnae.

SCHAPER, A.
1890.—Die Leberegelkrankheit der Haussäugethiere. Eine ätiologische und path-
ologisch-anatomische Untersuchung <Deut. Zeit. f. Thiermed., Bd. XVI,
pp. 1-95, Taf. I-V.
SCHÖNE.
(1886).—Beiträge zur statistik der Entozoen des Hundes. 8°. Inaug. Diss.
Leipzig.
SCHRANK, F. v. P.
1790.—Förtekning, på några hittils obeskrifne Intestinal-Kråk <Kongl. Veten-
skaps. Acad. nya Handl., XI, pp. 118-126.
SOMMER, H. O.
1896.—Results of an examination of fifty dogs, at Washington, D. C., for animal
parasites <Vet. Mag., III, p. 483-487.
SONSINO, P.
1876.—Intorno ad un nuovo parassito del bue (Bilharzia bovis) <Rendic. Accad.
Sc. Fis. Nat. Napoli, XV, pp. 84-87.
1890.—Studi e notizi elmintologiche <Proc. Verb. d. Soc. Tosc. di Sci. Nat., 4
maggio, 16 pp.
1896.—Varietà di Fasciola hepatica e confronti tra le diverse specie del genere
Fasciola, s. st. <Proc. Verb. Soc. Tosc. Sci. Nat., 3 maggio, pp. 112-116.
STILES, Ch. Wardell.
1892.—Notes on Parasites — 7: A word in regard to Dr. Francis Distomum texanicum <American Vet. Rev., XV, March, pp. 732-733.
1893.—(See Stiles & Hassall, 1893.)
1894-1895.—The Anatomy of the Large American Fluke (Fasciola magna) and a
comparison with other species of the genus Fasciola, s. st. Containing
also a list of the chief epizootics of Fascioliasis (Distomatosis) and a
Bibliography of Fasciola hepatica by Albert Hassall < The Journal
Comp. Med. and Vet. Arch. 1894, XV, pp. 161-178; 225-243, Pls. I-II,
figs. a-g in text; 299-313, Pls. III-IV; 407-417; 457-462; 1895, XVI, pp.
139-147; 213-222, Pls. VII-VIII; 277-282.
1895.—Notes on Parasites—32: On the rarity of Taenia solium in North America
< Vet. Mag., II (5), May, pp. 281-286.
1895.—Notes on Parasites—31: On the Presence of Adult Cestodes in Hogs < Vet.
Mag., II (4), April, pp. 220-222.
1896.—A Revision of the Adult Tapeworms of Hares and Rabbits < Proc. U. S.
Nat. Mus., XIX, pp. 145-235, Pls. V-XXV.
1897.—The Country Slaughterhouse as a Factor in the Spread of Disease < Yearbook of the Department of Agriculture for 1896, pp. 155-166.
———— & HASSALL, A.
1893.—A Revision of the Adult Cestodes of Cattle, Sheep, and Allied Animals
< Bulletin 4, Bureau of Animal Industry, U. S. Department of Agricul-
ture, pp. 1-134, Pls. I-XVI. Washington, D. C.
1896.—Notes on Parasites—44: Dicrocoelium lanceatum Stiles & Hassall, 1896.
< Vet. Mag., III (3), March, p. 158.
STOSSICH, M.
1892.—I distomi dei Mammiferi < Programma della civica scuola Reale superiore.
42 pp., Trieste.
TASCHENBERG, O.
1889.—Bibliotheca Zoologica, II, pp. VIII, 865-1730. Leipzig.
THOMAS, A. P.
1882.—Second Report of Experiments on the Development of the Liver Fluke
(Fasciola hepatica) < Journ. Roy. Agric. Soc. of England, XVIII, II, pp.
439-455, figs. 1-6.
1883.—The Natural History of the Liver Fluke and the Prevention of Rot.
< Journ. Roy. Agric. Soc. of England, XIX, I, pp. 276-305, figs. 1-20.

THOMAS, J. D.
   1884.—Hydatid disease, with special reference to its prevalence in Australia. 220
      pp., 5 pp. Adelaide.
WARD, H. B.
   1896.—A New Human Tapeworm (*Tænia confusa* n. sp.) < Western Med. Rev.,
      I., no. 2, pp. 35–36, figs. 1–2.
   1897.—Animal Parasites of Nebraska < Report Nebr. St. Bd. Agric. for 1896,
      pp. 173–189, figs. 1–12.
WEINLAND, D. F.
   1858.—Human Cestoides. An Essay on the Tapeworms of Man, giving a full
      account of their nature, organization, and embryonic development; the
      pathological symptoms they produce, and the remedies which have
      proved successful in modern practice. To which is added an appendix,
      containing a catalogue of all species of helminthes found in man. 8°.
      93 pp., 12 figs. Cambridge (Mass.). (Actual date of publication, prior
      to September 30, 1858.)
WERNICKE, R.
   1886.—Die Parasiten der Haustiere in Buenos Ayres < Deut. Zeit. f. Thiermed. u.
      vergl. Pathol., XII, p. 304.
ZEDER, J. G. H.
   1800.—Erster Nachtrag zur Naturgeschichte der Eingeweidewürmer mit Zusät-
      zen und Anmerkungen herausgegeben. 4°. 320 pp., Taf. I–VI. Leipzig.
   1803.—Anleitung zur Naturgeschichte der Eingeweidewürmer. 432 pp. 8°.
      Taf. I–IV. Bamberg.
ZÜRN, F. A.
   1882.—Die tierischen Parasiten auf und in dem Körper unserer Haussäugetiere.
      316 pp., Taf. I–IV. Weimar.

INDEX TO TECHNICAL NAMES.

[Synonyms in *italics* (*Acephalocystis ansa*); the more important references to each name in bold type (**113**).]

	Page.
Acephalocystis	113, 115
ansa	113
communis	113
cystifera	113
endogena	113, 116
eremita sterilis	113
exogena	113, 116
granosa	113
granulosa	113
humana	113
intersecta	113
macaci	113
ovis trayelaphi	113
ovoidea	113
piana	113
prolifera	113
prolifera socialis	113
racemosa	113
simplex	113
suilla	113
surculigera	113
Agamodistomum	28
suis	22, 28, 29, 143
Akis spiuosa	86
Alce alces	142
Alces machlis	142
palmatus	142
Alyselminthus expansus	128
Amabilia	69
Amphistoma	24, 64
bothriophorum	24, 67
cervi	24, 64, 65, 66, 67, 139, 140, 141, 142
conicum	64
crumeniferum	67
explanatum	24, 67, 140
tuberculatum	67, 140
Amphistomidae	22, 24, 27, 64
Amphistomum conicum	64
Anisolabis annulipes	86
Anoplocephala Vogti	127
Anoplocephalinae	24, 25, 68, 70, 125
Antilocapra americana	141
Antilope euchore	141
leucoryx	141
saiga	141
Asopia farinalis	86
Astoma acephalocystis	113
Atrypanorhyncha	68
Auchenia llama	143

	Page.
Bilhartzia crassa	60
Bilharzia	60
bovis	60
capensis	58
haematobia	58
crassa	60
hominis	59
magna	58
magna	58, 59, 138
Bos bubalis	23, 48, 139
frontalis	139
indicus	139
taurus	20, 23, 28, 48, 140
Boselaphus tragocamelus	141
Bothriocephalidae	84, 85
Bothriocephalinae	85
Bothriocephalus	85, 101, 103, 105
cordatus	85, 137, 138
cristatus	85
fuscus	138
latus	84, 85, 137, 138
Mansoni	85, 137
serratus	138
tropicus	72
Camelus bactrianus	143
dromedarius	143
Canis familiaris	138
Capra aries	112
hircus	141
pyrenaica	141
Capreolus caprea	142
Cariacus americanus	141
campestris	141
nambi	142
paludosus	142
rufus	142
simplicicornis	142
virginianus	142
Carnivora	138
Castor fiber	139
Cavia cobaya	139
Cercaria pigmentata	64, 65, 143
Cercopithecus cephus	138
cynosurus	138
fuliginosus	59, 138
mona	138
patas	138
sabaeus	138
Cervus *alces*	142

151

Page.

Cervus axis	142
canadensis	142
dama	142
elaphus	142
tarandus	142
unicolor	142
Cestoda	21, 24, 68
Cetacea	143
Cittotaenia denticulata	126
Cladocoelium giganteum	49, 51
hepaticum	29
Coenurus	11, 21, 25, 69, 70, 85
cerebralis	25, 70, 106, 108,
	109, 110, 111, 112, 139, 140, 141, 142, 143
serialis	110, 139
Coregonus albula	85
lavaretus	85
Cynocephalus maimon	139
Cysticercus	21, 24, 25, 69, 70, 73, 74, 85
acanthotrias	89
albopunctatus	89
bothryoides	89
bovis	11, 12, 17, 25, 70,
	71, 74, 75, 79, 80, 83, 92, 101, 140, 141, 143
caprinus	96
cellulosae	11, 12, 25, 70, 79, 83, 89, 90, 92,
	93, 94, 95, 101, 123, 137, 138, 139, 142, 143
cellulosus	89
clavatus	96
dicystus	89
echinococcus	113
finna	89
finnus	89
Fischerianus	89
inermis	71
lineatus	96
melanocephalus	89
multilocularis	89
ovis	96
pyriformis	89
racemosus	89
simiae	96
solium	69
suis	89
Taeniae mediocanellatae	71
saginatae	71
tenuicollis	11, 25,
	28, 70, 78, 79, 93, 96, 97, 101, 103,
	137, 138, 139, 140, 141, 142, 143
turbinatus	89
visceralis simiae	96
Cysticerkus bovis	71
cellulosae	89
tenuicollis	96
Davainea madagascariensis	86, 137
Dicotyle	50
Dicrocoelium	22, 55
lanceatum	23,
	55, 56, 137, 138, 139, 140, 141, 142, 143,
lanceolatum	55
pancreaticum	23, 55, 56, 57, 140
Digenea	27
Dipylidiinae	68, 85, 86
Dipylidium caninum	84,
	85, 86, 102, 103, 104, 105, 137, 138

Page.

Diskostoma acephalocystis	113
Distoma capense	58
(Cladocoelium) hepaticum	29
coelomaticum	57
(Dicrocoelium) coelomaticum	57
lanceolatum	55
hepaticum	29, 49
lanceolatum	55
pancreaticum	57
Distomum americanum	51
(Bilharzia) haematobium	60
caviae	48
crassum	51
(Fasciola) hepaticum	29
giganteum	49
haematobium	58
hepaticum	29
lanceolatum	55
magnum	49
musculorum suis	28
oculi-humani	48
ophthalmobium	48
pancreaticum	57
texanicum	51
Echinococcifer	70
echinococcus	114
Echinococcus	11,
	12, 21, 25, 69, 70, 79, 85, 101, 113, 118
altricipariens	113, 116
alveolaris	113, 117
arietis	113
cerebralis	113
cerebri	113
coenuroides	113
cysticus	113
endogena	113
giraffae	113
granulosus	113, 116
hepatis	113
hominis	113, 117
hydatidosus	113, 116
infusorium	113
intercranialis	113
lienis	113
mesenterii	113
multilocularis	113, 117
exulcerans	113
hepatis	113
multiplex	113
osteoklastes	113
pardi	113
polymorphus	25, 70, 79,
	113, 117, 137, 138, 139, 140, 141, 142, 143
process. vermiformis	113
pulmonum	113
racemosus	113, 117
retroperitonealis	113
scolicipariens	113, 116
simiae	113, 117
simplex	113, 116
subphrenicus	113
unilocularis	113
veterinorum	113, 117
Echinokokkus	113
Elephas indicus	139

Page.

Equus asinus 130
   caballus............................... 130
Fasciola.............................:..... 22, 27, 29
   americana 49
   Buchholzii 55
   carnosa 49
   cervi 64
   elaphi 64
   gigantea............................. 49
   gigantica..... 23, 29, 48, 49, 50, 51, 137, 140, 141
   hepatica............... 22,29, 30, 31, 32, 33, 34,
      35, 37, 38, 41, 42, 43, 45, 51, 52, 53, 55,
      56, 57, 137, 138, 139, 140, 141, 142, 143
   aegyptiaca 23, 48, 49, 50, 139, 140
   angusta 23, 48, 49, 137, 140
   caviae 48, 139
   humana 29
   Jacksoni............................. 29
   lanceolata 55
   magna 22, 27,
      29, 42, 49, 51, 52, 53, 54, 55, 140, 141, 142
Fasciolidae 22, 27, 28
Fasciolinae 22, 28
Fasciolaria hepatica..................... 29
Felis domestica 138
Festucaria cervi 64
Finna...................... 89
Gammarus Simoni 59
Gastrothylax........................... 24
   Cobboldii..................... 24, 67, 69, 139
   crumenifer 24, 67, 68, 69, 140
   crumeniferum..................... 67
   elongatum.................. 24, 67, 70, 139
   gregarius.................. 24, 67, 71, 139
Gazella dorcas 141
   euchore 141
Giraffa camelopardalis 141
Gynaecophorus bovis 60
   crassus............................ 60
   haematobius........................ 58
Halysis marginata 96
   ovina 128
   solium 71, 72, 90
Hexathrydium venarum 48
Hippotragus equinus 141
Homalogaster........................... 24
   paloniae 24, 67, 72, 139
   Poirieri.................... 24, 67, 140
Homo sapiens.......................... 137
Hydatides.............................. 96
Hydatigena cerebralis 109
   globosa 96
   granulosa 113
   oblonga 96
   orbicularis 96
Hydatis animata....................... 96
   erratica 113
   finna 89
   humana............................ 89
   piriformis......................... 89
Hydatula solitaria 96
Hydra hydatula........................ 96
Hymenolepis........................... 86
   diminuta..... .... 86, 137
   murina 86, 137

Page.

Inuus cynomolgus....................... 138
   ecaudatus 138
Kobus ellipsiprymnus.................... 141
Krabbea grandis........................ 85, 137
Lepus californicus 130
   callotis 130
   cuniculus 139
   domesticus......................... 130
   timidus 130
   variabilis.......................... 130
Limacidae.............................. 50
Limnaea 55, 56
   humilis 43
   oahuensis...................... 32, 43, 143
   peregra 42, 43
   rubella........................ 32, 43, 143
   truncatula 32, 42, 43, 46, 143
   viator............................. 43
Linguatula rhinaria.................. 119, 121
Lota lota.............................. 85
Lucius lucius 85
Lumbricus hydropicus 96
   latus 72
Macacus cynomolgus 138
   inuus 138
   silenus............................ 138
Macropus giganteus..................... 143
   major 143
Malacocotylea.......................... 27
Mammalia 138
Marsupialia............................ 143
Mazama rufus 142
Mesocestoides......................... 102
   lineatus 105, 138
Mollusca.............................. 143
Moniezia 26, 126, 127
   alba 26, 126, 127, 140
   dubia 127
   Benedeni 26, 126, 128, 140
   caprae 141
   denticulata 126
   expansa 26, 126, 127,
     128, 129, 130, 131, 132, 139, 140, 141, 142
   fimbriata.......................... 140
   Neumanni 26, 126, 128, 140
   nullicollis 26, 126, 140
   ovilla 129
   macilenta.................. 129, 130
   planissima... 26, 120, 121, 122, 126, 127, 132, 140
   trigonophora....... 27, 126, 128, 131, 132, 140
   Vogti.................. 26, 126, 127, 140
Monostoma conicum..................... 64
   elaphi............................. 64
Monostomum hepaticum suis............ 28, 96
   lentis 48
Multiceps 109
Mus rattus............................ 130
Nemathelminthes 21
Neotaenia 89
Onchorrhynchus Perryi................. 96
Orca gladiator......................... 143
Oryx beisa 141
   leucoryx 141
Ovibos moschatus...................... 140
Ovis ammon........................... 140

Page.

Ovis argali 140
  aries 20, 28, 140
  laticauda 141
  musimon 141
Ovuligera carpi 113
Papio maimon.....;...................... 138
Pentastoma coarctata 72
Perca fluviatilis 85
Phachochoerus aethiopicus 148
  africanus 143
Physa alexandrina 65, 143
  micropleura 65, 143
Planaria latiuscula 20
Planorbis 56
  marginatus 56
Plathelminthes 20, 21
Polycephalus bovinus 109
  echinococcus 113
  granosus 113
  granulosus 113
  hominis 113
  humanus 113
  ovinus 109
Potamochoerus penicillatus 143
  porcus 143
Pulex serraticeps 80
Pulmonata 56
Rodentia 139
Rupicapra tragus 141
Saiga tartarica.......................... 141
Salmo lacustris 85
  salar 85
  trutta 85
  umbla 85
Scaurus striatus......................... 86
Schistosoma................... 23, 27, 58, 63
  bovis 23, 58, 60, 61, 62, 140
  haematobium.... 23, 57, 58, 59, 60, 137, 138, 140
Schistosomum bovis....................... 60
  haematobium.......................... 58
Schistosominae................... 22, 23, 58
Sciurus cinereus 139
  vulgaris............................. 139
Semnopithecus entellus 138
Simia faunus............................ 137
  inuus 137
  rubra 137
  silenus 138
Solium ....:.... 72
Stilesia............. 26, 27, 127, 130
  centripunctata 27, 126, 130, 140
  globipunctata 27, 126, 130, 140
Strigea cervi............................ 64
Strongylus contortus............... 128, 132
Sus scrofa............................. 143
  domestica 20, 28, 143
Taenia............................ 24, 70
  abietina 72
  aculeata 120
  alba 127
  albopunctata 89
    hominis 89
  algérien 72
  algeriensis........................... 72
  apri.................................. 96

Page.

Taenia (Arhynchotaenia) echinococcus 114
  Benedeni 128
  bovina................................ 96
  Brandti.............................. 130
  capensis.............................. 72
  caprina 96
  cateniformis..................... 96, 114
    lupi 96
  cellulosae............................ 89
  centripunctata....................... 130
  cerebralis............................ 109
  coenurus 70, 96, 99, 102, 103,
        104, 105, 107, 108, 109, 110, 111, 112, 138
  confusa 85, 137
  continua 72, 90
  crucigera............................ 142
  cucumerina 114
  cucurbitina 71, 89
      grandis saginata 71
      pellucida 89
      plana pellucida.................. 89
      saginata 71
  (Cysticercus) acanthotrias 90
  (Cystotaenia) mediocanellata 72
      solium............................ 90
  degener.............................. 90
  de la première espèce................. 72
      seconde espèce 72
  dentata......................... 72, 90
  denticulata.......................... 126
  echinococca 114
  (Echinococcifer) echinococcus 114
  echinococcus 70, 85,
      102, 103, 104, 105, 113, 114, 115, 124, 138
  (Echinococcus) echinococcus 114
  echinokokkus 114
  ezpansa 127, 128
  fenestrata 72, 90
  ferarum.............................. 96
  fimbriata............................ 128
  finna 89
  flavopunctata 86, 137
  fusa............................. 72, 90
  Giardi 129
  globipunctata........................ 130
  globosa 96
  granulosa 113
  hamoloculata 90
  humana armata 90
  hydatigena 96
      anomala 89
  hydatigera 89
  hydatoidea 96
  inermis 72, 105
      fenestrata 72
  Krabbei...................... 102, 138
  lata 72
  longissima. 72
  lophosoma 72
  lupina............................... 96
  marginata 70, 89, 96,
        97, 98, 99, 100, 101, 102, 103, 104, 105, 138
  mediocancellata 72
  mediocanellata 72
      hominis 72

	Page.
Taenia *megaloon*	72
(*Moniezia*) *expansa*	128
planissima	127
trigonophora	128
mummificata	72
muscularis	80
nana	86, 114, 137
nigra	72
officinalis	90
ovilla	96, 126, 129
ovina	128
ovipunctata	130
pseudo-cucumerina	105
pyriformis	89
saginata	11, 68, 70, 71, 72, 73, 74, 75, 76, 77, 81, 83, 84, 85, 86, 87, 89, 94, 137
sans épine	72
scalariforme	90
secunda Plateri	72
serialis	102, 103, 104, 105, 110, 112, 138
serrata	98, 99, 102, 103, 104, 105, 138
serrata	114
simiae	96
solitaria	71, 90
solinum	11, 68, 70, 71, 83, 84, 85, 86, 89, 90, 91, 92, 94, 96, 137
"*solium* of dogs"	96
solium abietina	72
continua	72, 90
fenestrata	90
fusa	72, 90
mediocanellata	72
minor	72, 90
scalariforme	90
(*Stilesia*) *centripunctata*	130
tenella	90
tropica	72
turbinata	90

	Page.
Taenia *vervecina*	96
vesicularis	100
visceralis socialis granulosa	113
Vogti	127
vulgaris	72, 90
zittaviensis	72
Taeniarhynchus mediocanellata	72
Taeniidae	68, 84, 85, 101
Taeniinae	24, 68, 70, 83
Tarandus rangifer	142
Tenia armata umana	90
Ténia sans épine	72
Tetrassichiona	68
Thecosoma haematobium	58
Thymallus vulgaris	85
Thysanosoma	127, 128
actinioides	25, 126, 128, 133, 134, 136, 140, 142
Giardi	20, 126, 127, 129, 130, 140, 143
ovilla	130
ovillum	130
Tomiosoma	68
Trachélocampules	89
Trachelocampylus	89
Trematoda	21, 22, 27
Trichodectes canis	86
Ursus arctos	138
Vermes vesiculares	96
Vermis cucurbitinus	72
vesicularis eremita	96
socialis	100
Vesicaria finna suilla	89
granulosa	113
hygroma humana	89
lobata suilla	80
orbicularis	96
socialis	100
Vulpes lagopus	138

INDEX TO SUBJECTS.

	Page.
Abattoir inspection for—	
beef measles	77
gid bladder worm	112
hydatid disease	121
lancet fluke	57
large American liver fluke	55
ment of animals with flukes	47
pork measles	92
tapeworms in intestines of sheep and cattle	136
thin-necked bladder worm	101
Africa, eating of flukes by natives	66
American liver fluke—	
egg	52
large, disease	52
notes	49
pathology	53
position in host	54
prevalence and life history	51
relation to cattle industry	54
source of infection	55
Amphistomes—	
discussion of family	64-67
list of species	67
Areca nut, use in tapeworm disease of sheep and cattle	133
Arkansas, prevalence of fluke disease in cattle	51
Armed tapeworm, danger to man of infection with larvae	87
Australia, death among cattle from conical flukes	66
Beef measles—	
adult stage of tapeworm	71
destruction by cold storage	83
cooking meat	81
salt solution	17
discussion, and life history of tapeworm	71
disease in cattle	75
influence of age and sex on infection of cattle	80
season	80
position of parasites	78
prevention by process to kill parasites	78
in cattle	77
suggestions for diagnosis	77
symptoms in cattle	76
Beef, measly—	
manner of disposition	81
methods of preparation for food	82
prices in Berlin	82
Beef, prices of diseased grades in Saxony	20
Benzine, use for fluke disease in steers	45
Berlin, statistics of abattoirs for cysticercus in cattle	80
Bibliography of the more important works cited	145-150
Bilharziosis—	
disease in man from fluke	61
prognosis and treatment in man	63
Bladder—	
of man, effect of bilharziosis	62
worm, appearance of calcareous bodies	74
cysticercus, characters	71
destruction by salting	82
ease of recognition	78
gid, of sheep and calves, discussion	108-112
position in disease of beef measles	78
salting as means of destruction	82
thin-necked, life history	97
prevention of disease	101
Blood flukes in man and cattle—	
life history	58
probabilities of appearance in United States	64
Bluestone, use in tapeworm disease of sheep and cattle	133
Bothriocephalidae, description	85
Bovine blood fluke, discovery in Egypt	60
Bread cakes, kind used for fluke disease in sheep	44
Burial of diseased meats, grounds for opposition	16
Bunk on use of benzine for steers with fluke disease	45
Butchers, value of information regarding tapeworms	12
California, fluke disease among dairy cows	53
Calves—	
gid bladder worms, discussion	108-112
stages of disease with gid worm	110
treatment and prevention of gid disease	111
Cattle—	
adult tapeworms	125
and sheep, infestation with tapeworms	68
diagram showing season of danger from flukes	41
differential diagnosis of parasites	78
discussion of echinococcus hydatid	113-125
disease from common liver fluke	36
thin-necked bladder worm	99

Cattle—Continued.   Page.

effects of common liver fluke............ 34

  large American liver fluke......... 54

forms of *Moniezia*....................... 127

frequency of *Cysticercus bovis*.......... 80

general precautions against fluke dis-

  ease.................................... 47

hydatid disease.......................... 118

infection with beef measles 75

influence of age and sex on beef mea-

  sles infection........................... 80

  in disease from flukes................... 40

  season on beef measles 80

lancet fluke 55

lesions of bovine blood fluke 60

life history of adult tapeworm 130

location of beef measles................ 78

manner of receiving infection with

  tapeworms........................... 72

means of prevention of beef measles ... 77

  fluke disease 46

methods of preventing infection from

  tapeworm of man 83

pancreatic fluke, description 57

parasitic worms.......................... 20

post-mortem in beef measles........... 76

preparations for treatment of fluke dis-

  ease.................................... 44

prevention of tapeworm disease 135

snails as source of infection with flukes. 42

species of flukes........................ 28

suggestions for diagnosis of beef

  measles 77

symptoms of beef measles 76

table showing number condemned for

  hydatids............................. 123

tapeworm disease 131–136

tendency to fatten from fluke disease .. 45

treatment for tapeworm disease........ 133

  of disease from flukes.............. 43

  of verminous diseases............... 15

Cestodes or tapeworms—

discussion.............................. 68

order of flat worms, discussion 21

Cochin China, disease from pancreatic fluke. 57

Coenurus, characters....................... 70

Cold storage—

effect on diseased meats................ 17

use in rendering measly beef whole-

  some 83

Compendium of parasites arranged accord-

ing to their hosts.................... 137–143

Conical fluke of cattle and sheep, life his-

tory................................... 64

Cooking as means of—

killing beef measles.................... 81

making diseased meats safe for food.... 17

Copenhagen, statistics for beef measles.... 80

Curtice, statement as to treatment of adult

tapeworm in sheep.................. 129

Cysticerci, methods of killing.............. 81

Cysticercus—

characters 70

methods of finding in meat............. 79

time of development 73

Delafond, formulas for fluke disease........ 44

Dicrocoeles, species 55

Diet, kind necessary for man in treatment

  for tapeworm........................ 86

Dinwiddie on fluke disease in Arkansas ... 51

  post mortem examination for American

   liver fluke......................... 53

Dipylidinae, characters 86

Distomes, discussion 28

Dogs—

  adult tapeworm 123

  discussion of adult tapeworms....... 101–108

   echinococcus tapeworm 113–125

   gid tapeworm.................... 108–112

  life history of hydatid tapeworm...... 114

  management to prevent tapeworm dis-

   ease in man 125

  necessity for exclusion from slaughter-

   houses 15

  number and percentage infected with

   tapeworms 105

  prevention of tapeworm important to

   public hygiene and to farm profits... 11

  tapeworm disease 102

  treatment for tapeworms............. 106–108

Domesticated animals, treatment and pre-

vention of hydatid disease 121

Dose, size for sheep and cattle in tapeworm

disease 134

Echinococcus, characters................. 70

Echinococcus hydatid and tapeworm 113–125

Egypt, disease from blood fluke............ 60

Egyptian liver fluke, notes................. 48

Europe, benefit from inspection of meat for

"measles "............................ 89

European cities and towns, management of

slaughterhouse 14

Eye, decrease of disease as result of inspec-

tion 11

Fascioles, forms in American cattle........ 29

Feed, care for prevention of tapeworm dis-

ease 136

Fern—

  male, extract, use against tapeworm in

   man 88

  root, use in tapeworm disease of sheep

   and cattle 133

Fertilizer, use of diseased meats........... 16

Flatworms, two orders, discussion........ 21

Fluke disease—

general precautions 47

in animals, preventive measures........ 46

lack of laws in America.............. 48

preventive measures 46

Flukes—

and tapeworms of cattle, sheep, and

  hogs, key............................ 21

bovine, blood, freedom of man from

  danger of infection................... 64

common liver, diagnosis of disease in

  animals 39

diagrams illustrative of occurrence. 41

effects on cattle, sheep, and hogs.... 34

generations 33

in man 48

names of disease.................... 34

Flukes—Continued.   Page.

common liver of cattle, sheep, and hogs.   29
    time of infection of cattle..........   42
    varieties............................   48
    conical, distribution...................   65
    geographical distribution and seasons .   40
    in cattle, sheep, and hogs, key...... ...   22
    means for prevention of scattering eggs
       in fields   46
    of cattle, pancreatic.....................   57
    or trematodes, technical discussion   27
    order of flat worms, discussion.........   21
    pathological effects upon domesticated
       animals   36
    position in diseased animals   40
    snails as source of infection of cattle ..   42
    source of infection of man with bilhar-
       ziosis................................   61
    treatment of disease of cattle   43
Francis, report on fluke disease 53, 54
Freibank, history of German system of sell-
    ing inferior meats.....................   19
Fringed tapeworm—
    cause of disease in sheep................   128
    treatment of disease in sheep..........   129
Germany—
    effect of meat inspection on human
       health................................   11
    frequency of hydatid disease   122
    prices of measly beef....................   83
    system of selling diseased meats.......   18
Giant liver fluke, notes   49
Gid bladder worm—
    disease, treatment and prevention for
       calves and sheep.....................   111
    in sheep and calves   108-112
    tapeworm in dogs, discussion........   108-112
Hard-shell tapeworms, description and clas-
    sification............................   70
Hauber's lick for sheep with fluke disease.   44
Heat, temperature necessary to destroy
    bladder worms........................   82
Hertwig, note on destruction of beef mea-
    sles by heat..........................   82
Highlands, freedom from flukes............   40
Hogs—
    diagram showing season of danger from
       flukes   41
    disease from common liver fluke   36
       thin-necked bladder worm   99
    of measles and position of parasites   92
    effects of common liver fluke...........   34
    general precautions against fluke dis-
       ease..................................   47
    guaranty of freedom from infection with
       pork measles   91
    infection with tapeworms   68
    lancet fluke   55
    means of prevention of fluke disease....   46
    parasite worms.........................   20
    pathology of hydatid disease   120
    Prussian statistics of frequency of *Cys-*
       ticercus cellulosae   93
    raising at slaughterhouses   15
    species of flukes   28
    treatment for verminous diseases   15

Human beings—   Page.

blood fluke, life history   58
prevention of spread of tapeworms.....   11
Hutcheon, Dr., method of treating tape-
    worm disease   133
Hydatid—
    cysts, modifications....................   114
    disease in—
       hogs   120
       man..................................   124
       prevention in man   125
       sheep   119
       various animals   117
    frequency in various animals   121
    in man and domesticated animals, dis-
       cussion 113-125
    tables showing cases by nationality,
       age, and localities 124, 125
    tapeworm, life history..................   114
Iceland, frequency of hydatid disease......   122
India, frequency of hydatids in domesti-
    cated animals   122
Inspection—
    method for beef measles.................   78
    of abattoirs for beef measles   77
       meat of animals diseased with flukes   47
Inspectors, meat, necessity for information
    about tapeworms......................   11
Japan, disease from pancreatic fluke.......   57
Jurisprudence in regard to fluke disease ...   48
Kamala, use in tapeworm disease of sheep
    and cattle   133
Kidney of man, bilharziosis   62
Lambs--
    effect of tapeworm disease   128
    stages of disease from gid worm   110
Lancet fluke in cattle—
    life history.............................   56
    sheep, and hogs   55
Laws regarding fluke disease   48
Leuckart, chart for fluke disease in cattle
    slaughtered in Berlin   40
Lewis, investigation of effects of heat on
    bladder worms........................   82
Liver—
    echinococcus, symptoms   119
    fluke, common, effects on cattle, sheep,
       and hogs   34
       in man   48
       life history   30
       narrow, notes   48
       of cattle, sheep, and hogs...........   29
       varieties   48
Man—
    adult and larval tapeworm   94
    common liver fluke.....................   48
    danger from larvae of armed tapeworm.   87
    diet in treatment for tapeworm........   88
    discussion of echinococcus hydatid .. 113-125
       tapeworm............. 82-101
    guaranty of freedom from infection with
       pork measles.........................   91
    hydatid disease   124
    infection with *Cysticercus cellulosae*   95
    methods for prevention of tapeworm...   89
    prevention of hydatid disease..........   125

Man—Continued. Page.
symptoms, diagnosis, and treatment of
infection with tapeworm............. 87
symptoms of bilharziosis............... 61
ways of determining species of tapeworm 86
Marginate tapeworm—
discussion 96
in dogs........................... 101–108
life history......................... 97
Market, question of sale of measly beef.... 81
Marshes, favorableness to disease from
flukes............................... 40
Marshy ground, methods of prevention of
fluke disease....................... 46
Measles—
beef, appearance of feverishness in
cattle............................. 75
life history of tapeworm............. 72
Measly beef—
manner of disposition.............. 81
rule of Ostertag, for salting for food... 83
Meat inspectors, necessity for information
regarding tapeworms.............. 11
Meats—
condemned, disposal............... 15
diseased, reasons against burial or burning............................... 16
disposal when affected with fluke disease............................. 47
infected, selling under declaration..... 18
infected with beef measles, question of
use............................. 78
object of report on flukes and tapeworms........................... 11
Mojowski, treatment of sheep with naphthaline for fluke disease.............. 45
Moniezia, genus of tapeworms......... 127
Monkey, sooty, parasite found by Cobbold. 59
Muscle fluke of swine................. 28
Mutton, prices of diseased grades in Saxony 20
Naphthaline, use for fluke disease in sheep. 45
Nomenclature, scientific, use in report..... 13
Ostertag—
compilation of data concerning measly
beef............................. 20
rule for cooking meat to kill beef
measles......................... 82
salting measly beef.............. 83
statement as to common liver fluke in
European cattle................. 36
Overdose in treatment of tapeworm disease,
management..................... 135
Parasites—
general methods of prevention of diseases........................... 14
of bilharziosis, pathology.......... 62
position......................... 61
position of flukes in diseased animals.. 40
worms in cattle, sheep, and hogs....... 20
Picric acid, use for tapeworm disease in cattle and sheep................... 133
Pillizzari, investigation of bladder worms.. 82
Pleuro-pneumonia, feature distinguishing
from hydatid disease............. 119

Pork— Pag
bladder worm, life history.............
measle tapeworm, prevalence in man,
and characters....................
measles, disease in hogs, inspection ....
measly, disposition...................
Preventive measures against fluke disease.
Prussia, frequency of infection of cattle
with cysticercus..................
Pumpkin seed—
for tapeworm in sheep.............. 1
use against tapeworm in man..........
Rats, necessity for exclusion from slaughterhouses........................
Rissling, method of finding cysticercus in
meat.............................
Salting as—
means of making diseased meat safe for
food...........................
method of destruction of bladder
worms in meats..................
Sandwich Islands, prevalence of fasciola-
sis..............................
Saxony, Kingdom, statistics of meat classification.........................
Schmidt-Mülheim, method of finding cysticercus in meat...................
Sicily, prevalence of fluke disease among
sheep...........................
Sheep—
adult tapeworms................... 1
danger to cattle from fluke disease.....
diagram showing season of danger from
flukes..........................
discussion of echinococcus hydatid.. 113–1
disease caused by fringed tapeworm... 1
from common liver fluke..........
from thin-necked bladder worm....
effects of common liver fluke..........
forms of *Moniezia*................. 1
frequency of lancet fluke..............
general precautions against fluke disease...........................
gid bladder worm, discussion........ 108–1
hydatid disease.................... 1
lancet fluke......................
lesions of bovine blood fluke..........
life history of adult tapeworm......... 1
means of prevention of fluke disease...
parasitic worms..................
prevention of tapeworm disease....... 1
species of flukes..................
tendency to fatten from fluke disease ..
treatment and prevention of gid disease........................... 1
for tapeworm disease.............. 1
of disease from flukes.............
fringed tapeworm.............. 1
verminous diseases..............
Slaughterhouse—
care to prevent tapeworm infection.... 1
125, 1
disposal upon abandonment...........
raising of hogs in yards...............
sanitary supervision ....
steps for segregation..............

ils as— .Page.
enemies of stock raisers.................. 42
means of destroying, for prevention of
fluke disease 47
source of infection with large American
liver fluke............................ 55
ck raisers—
snails as enemies...................... 42
value of information regarding tapeworms 12
z, case of bilharziosis 61
ne. (See also Hogs.)
adult tapeworms........................ 126
discussion of echinococcus hydatid .. 113-125
nia marginata in dogs, period of devel
opment............................... 104
niidae, description...................... 85
discussion of family of tapeworms.... 68-136
eworm disease—
caution in treatment 134
dogs 102
in cattle and sheep..........,....... 131-136
man, decrease as result of inspection of meats.................... 11
sheep and cattle, prevention 135
eworms—
adult, in cattle and sheep.............. 125
and flukes of cattle, sheep, and hogs, key. 21
differences of cysticercus and echinococcus.................................. 79
dogs 123
fringed, cause of disease in sheep...... 128
treatment of disease in sheep 129
gid, in dogs, discussion.............. 108-112
hard shell, description and classification 70
life history.............................. 69
in cattle, sheep, and hogs, key 24
life history............................... 130
man, adult and larval 94
ways of determining species 86
marginate, characters......... 96
of dogs, key............................ 101
treatment of disease 106-108

Tapeworms—Continued. Page.
methods for prevention in man......... 89
number and percentage of infection of
dogs 105
of man, key 84
methods of preventing infection of
cattle 83
of Taeniidae family, characters 68
or cestodes, discussion 68-136
order of flat worms, discussion......... 21
hogs 92
symptoms, diagnosis, and treatment of
infection in man................... 87
of pork measles, inspection 92
time of development of cysticercus.... 73
transmissibility from animals to man .. 68
unarmed distribution................... 84
Temperature necessary to destroy bladder
worms 82
Texas—
outbreak of fluke disease 53
prevalence of fluke disease among cattle.................................... 42
Thysanosoma, genus of tapeworms....... 128-130
Tonkin, disease from pancreatic fluke...... 57
Trematodes or flukes—
order of flat worms, discussion......... 21
technical discussion.................... 27
Trichinous hogs, disinterment and eating of
carcasses buried by sanitary officials. 16
Turpentine, use against tapeworm......... 88
United States, frequency of hydatid dis-
ease................................. 122
Virchow on proportion of cysticercus in
man................................. 95
Water—
care for prevention of tapeworm disease. 136
danger in districts infected with bil-
harziosis 63
Wet years, effect on flukes in animals...... 40
Wool, effect of tapeworm disease........... 128
Worms, parasitic, of cattle, sheep, and hogs. 20
Zündel, division of periods of disease from
common liver fluke 36

www.ingramcontent.com/pod-product-compliance
Lightning Source LLC
Chambersburg PA
CBHW021810190326
41518CB00007B/532